D1213788

# RFID in the Supply Chain

## About the Author

**Pedro M. Reyes, Ph.D.,** is an associate professor in the Hankamer School of Business, director of the Center for Excellence in Supply Chain Management, and Lawrence Schkade fellow, Baylor University. His research has been published in several journals, including *International Journal of Integrated Supply Management*, *International Journal of Operations & Production Management*, and *International Journal of Data Analysis and Information Systems*. Dr. Reyes holds an annual symposium on RFID at Baylor University. Prior to his career in academia, he worked in operations management for more than 20 years.

# RFID in the Supply Chain

Pedro M. Reyes

New York Chicago San Francisco
Lisbon London Madrid Mexico City
Milan New Delhi San Juan
Seoul Singapore Sydney Toronto

The McGraw-Hill Companies

**Library of Congress Cataloging-in-Publication Data**

Reyes, Pedro M.
RFID in the supply chain / Pedro M. Reyes McGraw-Hill.—1st ed.
p. cm.
Includes index.
ISBN 978-0-07-163497-7 (alk. paper)
1. Business logistics. 2. Inventory control—Automation. 3. Radio frequency identification systems. I. Title.
HD38.5.R484 2011
658.7'87—dc22

2010045510

McGraw-Hill books are available at special quantity discounts to use as premiums and sales promotions, or for use in corporate training programs. To contact a representative please e-mail us at bulksales@mcgraw-hill.com.

**RFID in the Supply Chain**



1 2 3 4 5 6 7 8 9 0 DOC/DOC 1 9 8 7 6 5 4 3 2 1

ISBN 978-0-07-163497-7
MHID 0-07-163497-5

The pages within this book were printed on acid-free paper.

**Sponsoring Editor**
Michael Penn

**Acquisitions Coordinator**
Michael Mulcahy

**Editorial Supervisor**
David E. Fogarty

**Project Manager**
Manisha Singh, Glyph International

**Copy Editor**
Anita Wagner

**Proofreader**
Manish Tiwari, Glyph International

**Indexer**
Robert Swanson

**Production Supervisor**
Richard C. Ruzycka

**Composition**
Glyph International

**Art Director, Cover**
Jeff Weeks

# Contents

# Foreword

Let me take this opportunity to express my highest regard for Dr. Pedro Reyes. I have known Dr. Reyes for the past 20 years, as a student in my classes, as a professional in industry, and as a knowledgeable and well-respected member of academia. He has made an enormous impact on the lives of those he has worked with and the students he has mentored.

Pedro Reyes has drawn from his industry and academic experience to write a practical guide to RFID in the supply chain. Dr. Reyes has over 20 years of industry experience and over 8 years in academia, working in the area of supply chain management. He has more than 25 publications in academic journals and numerous academic awards, has served on eight journal editorial boards, and is an active consultant in the field, all of which reflect the reputation and respect he enjoys in the field of supply chain management in both industry and academia.

For the past five years, Dr. Reyes has held an annual symposium on RFID at Baylor University. These continuing symposiums receive support from industry, Sloan Industry Studies, and Baylor University. Through these symposiums, academia and industry have partnered in expanding the knowledge and application of RFID.

Through his unique style of writing and teaching, Dr. Reyes conveys his knowledge on the subject of technology and the supply chain. His practical experience is instrumental in making this book a must-read for those responsible for the movement of goods and delivery of services in a global economy.

In the first part of the book, Dr. Reyes elaborates on RFID from historical and technical perspectives. He gives the advantages and limitations of the technology and examines the standards for RFID.

In the second part of the book, Dr. Reyes discusses the challenges faced in implementation of RFID. He gives an overview of system architecture, variables, and factors that go into RFID implementation. Applications, security, and privacy issues are included in this part of the book. He concludes this section with the return on investment (ROI) aspects of RFID for business.

In the third section of the book, he summarizes case study examples of how RFID has been and is being used to improve supply chain visibility, asset visibility, and tracking of work in progress as well as returnable asset tracking. Dr. Reyes concludes the book with a look to the future for RFID and supply chain management.

For supply chain managers and professionals looking to learn more about RFID and its applications, this book offers a great learning experience. For academic and industry professionals, this book provides a deeper understanding of how RFID can improve supply chain management.

Patrick Jaska, PhD
*Professor of Business Systems*
*Chair, Department of Business Computer Information Systems*
*College of Business*
*University of Mary Hardin-Baylor*

# Preface

In the decades before my current position as a college professor, I witnessed firsthand many technologies that changed how businesses managed their internal operations as well as how these technologies changed the supply chain landscape. During those years, I mostly worked with replenishment systems. So when I was working on my dissertation, it made sense to study replenishment systems. I focused on grocery supply chains.

During the last few months prior to my dissertation defense, I was continuing to read various papers on supply chain management technologies. I read a short article on RFID, but really did not see how the technology could work. And since I was studying "business," I was not very interested in the engineering side of the technologies.

Well, it so happened that my replenishment system model was not getting the expected results. So I contacted the grocery retail store managers and asked for additional site visits to help me figure out what parameters (if any) might be incorrectly represented in my model. As I was entering one particular store in the Dallas area, the alarm went off (like someone had stolen something).

The funny thing was that the alarm went off as I was walking in (not out). As it turned out, the store was part of a pilot-study with Gillette on the use of RFID. Of course I did not have any of those products on me. So the questioning began. As it was discovered, I had an RFID tag in my shoe from a recent purchase. At this point, the wheels began to turn in my head, thinking, "How can this RFID tag be used to improve replenishment systems?" (This was in 2002, before Wal-Mart announced its mandate.)

Hence, my story is that I literally walked into the RFID field. The rest is history. I have enjoyed studying the various RFID applications and how this old/new technology has been reshaping the landscape of supply chain management processes.

I begin most of my professional talks by saying, "RFID has, for the most part, been flying below the business innovation radar." This has been my opening tag line dating back to my early work in 2002. Of course in those first few years of my studies I was questioned about return on investment (ROI), where's the business case, and standards. Hence, the real challenge during my academic or professional presentations has been to get others to look beyond the present cost. I explain that RFID is a proven technology with its roots in access control, and in theory there is clearly a great potential to revolutionize business. With the assumption that the engineering side of the RFID field can make it cost effective, how can we in business adopt the technology, reduce operating costs, and ultimately make more money?

With the publication of this book, I would like to acknowledge a number of colleagues and coresearchers who contributed to this work. The nature of my academic career depends heavily on collaboration. And over the years, my collaborators have also become my friends.

Patrick Jaska, professor at the University of Mary Hardin-Baylor, is a long-time colleague, coauthor, mentor, and friend. He has been a big supporter of my work and always willing to help.

For his encouragement when I was just starting to research RFID, Frank Giarratani, director at the University of Pittsburgh's Center for Industry Studies (http://www.industrystudies.pitt.edu), deserves a lot of credit. His support of my research based on an understanding of the RFID industry and the annual RFID Integrated Supply Chain Seminar Series/Symposium was instrumental in advancing my career toward tenure. And most recently, with his guidance and the Industry Studies Association (http://www.industrystudies.org) support, I coordinate an RFID Research Network. The network members are professional colleagues and friends who have generously shared ideas and questions over the past six years. They include Kevin Berisso, director for the Automatic Identification and Data Capture (AIDC) Laboratory at Ohio University (http://www.ohio.edu/industrialtech/aidc); Qiannong Gu, assistant professor of operations management at Sam Houston State University; Gregory Heim, assistant professor at Texas A&M University; Diego Klabjan, associate professor at Northwestern University; John Visich, associate professor at Bryant University; and Pamela Zelbest, assistant professor of operations management and director of the Sower Business Technology Laboratory at Sam Houston State University (http://www.shsu.edu/~coba/sower/).

I would also like to recognize Greg Frazier, professor of operations management (my dissertation chairman), and Edmund Prater, associate professor (also on my dissertation committee), both at the University of Texas at Arlington, who through their insights continue to ask challenging questions as they pertain to my research.

In addition, Nicole DeHoratius, assistant professor at the University of Portland, always had some interesting questions that pushed my research for a better understanding of RFID's potential.

Ertunga Ozelkan, assistant professor at the University of North Carolina at Charlotte, not only participated in the annual RFID Integrated Supply Chain Management Symposium but also helped organize special topics on RFID in supply chains at the Decision Sciences Institute (DSI), the Production and Operations Management (POMS), and the Institute for Operations Research and Management Sciences (INFORMS) annual meetings.

I also express my gratitude to those who participated and/or attended those special topics on RFID in supply chains. Without your feedback and questions, parts of this book would not have been possible.

I would also like to acknowledge three postgraduate assistants who took an interest in my RFID research and helped in data collection for various projects, including this book. These students include Sushmi Chakraborty (MBA/MSIS 2011), Christopher Zane (MBA 2008), and Manuja Baral (MS 2006).

In addition to my professional colleagues, my family and friends played an important role in encouragement and support. Friends kept asking for updates on the writing and would state: "You are the book." And finally a special thanks to my best friend and wife, Cherylle, who truly made a difference and continues to provide encouragement and inspiration.

PEDRO M. REYES

# PART I

# Introduction and Overview



The first part consists of three chapters and provides an overview for RFID. Chapter 1: *Introduction* first introduces the reader to RFID and why this old technology is receiving so much attention. Chapter 2: *RFID 101* provides a technical overview of the basics of RFID technology written in non-engineering speak. Chapter 3: *EPCglobal Overview and Standards* describes the three-layer architecture of the EPCglobal standards (identify, capture, and exchange), and is perhaps the most technical chapter of this book.

# CHAPTER 1
# Introduction

This chapter begins with a brief history of past business technologies, followed by an introduction to RFID and an explanation of why it is getting so much hype. Finally, the chapter ends with the motivation and organization of this book.

## Brief History of Past Business Technologies

Throughout history, business technologies have revolutionized the way firms design (and often redesign) their supply chains and management control systems. Historical examples include (1) the telegraph once used for railroad transportation scheduling, (2) the telephone (and facsimile) for faster business communication, and (3) electronic data interchange (EDI) for more efficient and paperless business transactions. While these technologies have enhanced the business practices of those eras, the actual benefits were limited to the specific supply chain process. The Internet has helped to address those limitations and the trade-offs between cost, rich content of data, real-time information sharing, and the up- and downstream integration between the business partners.

Today, RFID (radio frequency identification) is quickly catching on as an intriguing supply chain technology with flexibility for numerous applications. Despite its increasing popularity, the lack of understanding of RFID technology has slowed its acceptance. Yet, even with a limited and fragmented understanding, there is still a strong interest in RFID technology as a viable solution for improving supply chain operations.

## What Is This Thing Called RFID?

RFID is an auto-ID technology that uses radio frequencies to identify, track, and trace an object or product. Like many modern technologies, RFID technology has its origin in military applications during World War II (Fig. 1.1), when British planes were equipped with radio frequency transmitters to identify them as friendly aircraft to British forces on the ground. Commercial applications began during the early 1980s. Today, these applications span several industries.

**FIGURE 1.1** Origin in military applications. (*Source*: http://d2eosjbgw49cu5.cloudfront.net/rfid-weblog.com/imgname-rfid_and_its_history--50226711-36935578.jpg)

While there are arguments both pro and con, RFID has the potential to offer considerable benefits. A variety of applications already exist for RFID with over 100 reported cases in the fields of security, process control, hospital, consumer goods, retailing, document management, perishable logistics, warehousing, distribution, and construction sites. As more companies consider the potential applications of RFID, a good understanding is needed of what RFID is, the current and future states of RFID technology, and the current and future applications of RFID, as well as the technology's advantages and limitations.

## Why All the Hype?

RFID has, for the most part, been flying below the business-innovation and best practice radar. Most of the propaganda and press given to RFID have been since the mandates announced by Wal-Mart* in 2004 and the U.S. Department of Defense (DoD) in 2003 for suppliers' use of RFID, January 2005 is considered by researchers as the "big bang" for RFID. Yet, whether RFID represents a new direction in supply chain management theory and practice is a question

*Branded as Walmart since 2008.

of no small consequence. Equally important, it is not reasonable to believe that all firms will adopt RFID, yet many managers are in a dilemma as to whether RFID is right for their organization or application. In some ways, RFID is like many other past technological implementations, but in some ways it is not. The actual benefits and risks of RFID coupled with managers' evolving perceptions about these benefits and risks will decide the speed at which RFID moves from introduction and developmental stage to the maturity stage. Many RFID white papers published during the past few years describe RFID and its advantages, primarily to aid managers in their effort to determine whether RFID is appropriate for their particular needs and give them some guidelines for implementing an RFID solution. Although RFID has been around for more than 60 years, it took the recent mandates by Wal-Mart and the DoD to spark the massive interest in its potential for improving supply chain performance. Also, contributing to this interest is the rapid acceleration and availability of computer science and Internet technologies that have been evolving and reshaping supply chain management processes and practice. As part of the considerations for RFID implementation, managers must filter out the hype and understand what the technology can do—and equally important what it cannot do. As with many technologies, the excitement and the misunderstanding can be damaging to expectations.

## Motivation and Organization of this Book

This book is targeted to the business community and aims to enlighten and inform managers about RFID issues and the design principles behind software applications, and ultimately to help them decide whether, when, and how to use RFID technology for improving supply chain operations.

### Basics of RFID

The basics of RFID are presented in Chap. 2: *RFID 101*. The chapter begins with a history of RFID followed by an overview of RFID system components. The technology's advantages and limitations are also provided. Typically, the basic RFID system consists of tags, antennas, readers, and communication infrastructure (Figs. 1.2 and 1.3). Naturally an RFID tag is attached to an object. The RFID tag can be either active (battery powered and proactively emitting a radio frequency signal) or passive (unpowered and reactively emitting a radio frequency signal). An RFID reader communicates with the tag to identify the object to which the transponder or tag is attached. The tag has information about the object for identification purposes depending upon the need. The serial number, model number, or other characteristics of the object could be stored within the tag to identify

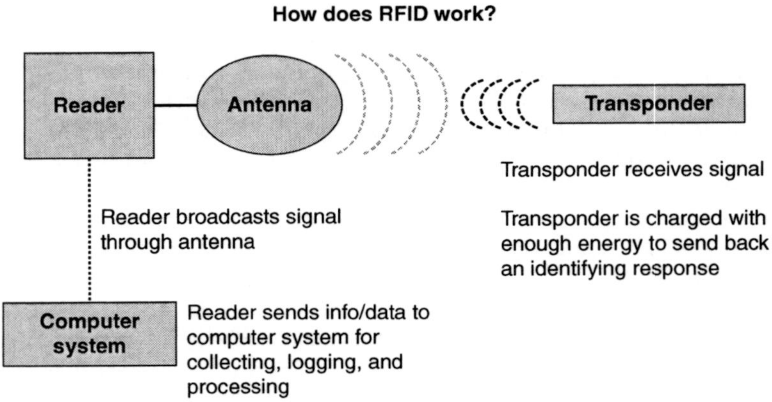

How does RFID work?
Reader
Antenna
Transponder
Transponder receives signal
Reader broadcasts signal through antenna
Transponder is charged with enough energy to send back an identifying response
Computer system
Reader sends info/data to computer system for collecting, logging, and processing

**Figure 1.2** Basic RFID system. (*Source*: http://www.traze.in/images_default2/about_rfid_img1.gif)

SMS or HITPS over GSM/GPRS
Operator network GSM/GPRS
RFID reader inbuilt in mobile phone
Centralized database
Wi-Fi AP/RFID reader
Server application
Internet
Backend systems

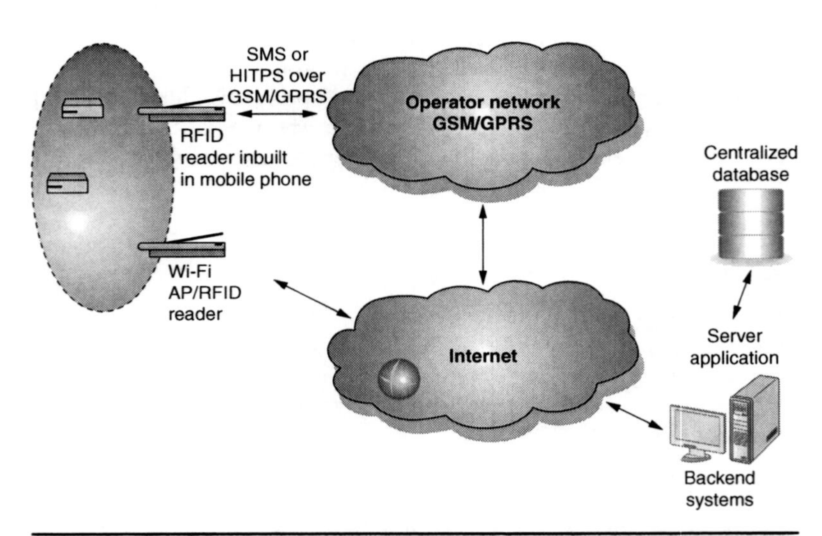

**Figure 1.3** RFID technology. (*Source*: http://www.traze.in/images_default2/about_rfid_img1.gif)

or distinguish this object from others or to allow tracking of the movement of the object.

## EPCglobal

An Electronic Product Code (EPC) is an identification scheme for the universal identification of physical objects via RFID tags and other means. The EPC components consist of an EPC Manager Number, an object class identification, and a serial number used to uniquely identify the instance of the object.

The structure of EPC tags was first developed at the Auto-ID Center at the Massachusetts Institute of Technology as "the internet of things" and is now managed by a not-for-profit joint venture between EAN International and the Uniform Code Council known as EPCglobal (see www.epcglobalinc.org and www.epcglobalus.org). This body manages the UPC (Universal Product Code) information found in bar codes and also sets the standards for how basic product information is encoded in RFID chips as well as how information is passed from RFID readers to various applications, including application to application. Chapter 3: *EPCglobal Overview and Standards* provides a comprehensive discussion of the EPCglobal framework and its three-layer architecture for the EPCglobal standards (identify, capture, and exchange).

## Challenges and Issues

Perhaps the leading challenge to designing an RFID application is achieving seamless integration and building consensus about the RFID strategy across the supply chain. First, it is difficult to design the application within a single firm (regardless of size) and then to extend it across a supply chain network of linked firms. Success has often been associated with the stronger channel leaders who are close to the ultimate customer and who have the buying power to influence the pace and direction of technology investment upstream in the supply chain. The challenge of building consensus about RFID adoption is a major topic in Chap. 4: *Challenges in Designing RFID Applications.*

Issues that have surfaced regarding RFID include security and privacy. Again, these issues are certainly not new to decisions on adopting business technology and should be at the forefront of any RFID considerations. While the public perception on security and privacy has some weight in the decision, it is mostly the organization's data security policies that must be examined to ensure customer data is not compromised. For the supply chain, security policies are outlined in EPCglobal Network measures set forth by EPCglobal (an international RFID standards body). Privacy advocates are more concerned about customers being tracked and their buying behavior monitored. Several solutions have been proposed to eliminate tracking of tags after products are sold. These solutions include "kill tags," password lock, cage approach, active jamming, and cryptography, and are presented in Chap. 5: *RFID Security and Privacy.*

A primary barrier to RFID adoption that is at the forefront of managerial concern is the difficulty of quantifying the cost-benefit ROI (return on investment) for acquiring this technology. While cost-benefit analysis is an ongoing business decision tool, many factors contributing to decisions on RFID adoption are similar to those involved in deciding whether to adopt the recent Internet-based e-commerce technologies. This adoption issue (which I call "show me the money") will be discussed in more detail in Chap. 6: *Business Analytics.*

## Case Studies

RFID has received increased attention from practitioners and academics, largely due to the mandates from Wal-Mart and the U.S. Department of Defense. In fact, January 2005 can be considered as the "big bang" for RFID. Other early adopters of RFID include Target, Metro Group, and Tesco. In addition, the U.S. Food and Drug Administration (FDA) strongly recommended that the pharmaceutical and health care industries adopt RFID as a way to protect the drug supply chain from counterfeiting and terrorist actions.

Wal-Mart's objective was to replace bar codes and scanners with RFID tags and readers in order to increase speed, efficiency, and security in the supply chain, and to reduce inventory levels, out-of-stock merchandise, and labor cost within the retail stores and warehouses. Other potential benefits in supply chain performance include (1) improved accuracy and security of information sharing across the supply chain; (2) reduced storage, handling, and distribution expenses; (3) increased sales through the reduction in out-of-stock merchandise; (4) improved cash flow through increased inventory turns and improved utilization of assets, and improved customer service and satisfaction; and (5) increased collaboration and planning across supply chain partners.

Such inventory management is one use of RFID. Other potential advantages are in homeland security, allowing agencies to screen people and materials as they pass through an airport, harbor, or any type of checkpoint. In health care, RFID can be applied in hospital settings to ensure that patients are not given new medication that interacts with other drugs already taken. The pharmaceutical industry can use RFID to resolve issues of counterfeiting and diversion of goods. RFID can be used to track assets from secure computers to priceless artwork. In courts of law, there are applications both to help manage the thousands of documents lawyers use to build a case and to protect the integrity of a chain of evidence for officers of the court. In libraries, RFID has reduced costs associated with lost inventory, as we would expect. Now a library can take a complete inventory of its holdings in a few days rather than months, and patrons can passively check books in and out by merely passing by a reader.

A broad range of RFID applications is now evident. Within the supply chain spectrum, one can observe applications for improving inventory management and controlling supply chain operations. Case studies in this book are used to illustrate these broad ranges of RFID applications across industries.

Improved supply chain visibility is one of the benefits of RFID. Chapter 7: *Supply Chain Visibility* demonstrates the applications found in the retail and pharmaceutical supply chains.

Another benefit of RFID is asset visibility in the supply chain. Chapter 8: *Asset Visibility* examines how RFID is used in hospitals and health care, as well as the improvements in capital goods tracking.

Chapter 9: *Work-in-Progress Tracking* describes how firms are using RFID to manage their internal supply chains. Internal benefits are illustrated as firms are able to improve inventory management from the time of ordering through final inspection before shipment.

A potential growth area for RFID is in libraries due to decreasing library budgets and decreases in the purchasing power of the customers. In a typical inventory management system, the RFID tag is used as a "throw-away" technology; as soon as the item reaches the customer, its life cycle ends. However, the use of RFID to identify a library book allows the tag to be used multiple times since the same book remains in circulation for a long period of time (i.e., it is checked in and out a number of times). Chapter 10: *Library Management System* provides examples of the benefits, such as locating missing books and identifying books that have been misplaced on the shelf.

Reusable assets are often misplaced, and the lack of visibility of their movement can lead to substantial losses for companies. Chapter 11: *Returnable Asset Tracking* illustrates how the RFID asset-tracking solution enables companies to better manage their returnable assets. This RFID application also provides information about assets due back from various trading partners, and presents information regarding the status of returnable assets against associated order numbers, improving the visibility of assets possessed by different partners across the supply chain.

## Summary and a Look Ahead

Ever since the Wal-Mart and Department of Defense decisions to move RFID to the forefront as a strategy for improving supply chain operations, RFID technologies and applications have matured quickly. As a result, RFID applications have become more efficient and effective tools for improving supply chain operational efficiency and enhancing customer service. Chapter 12: *Summing It Up and Looking Ahead* provides a summary of this book and a suggested blueprint for looking beyond your own benefits.

The RFID benefits and the associated costs with the price of readers and the potential impact on a firm's information technology (IT) infrastructure have been discussed. Although these are real costs that have affected the decision to deploy RFID systems, an important consideration is being overlooked and often neglected. For this technology to really make a difference, the benefits should be realized systemwide closed-loop with trading partners and the value gained should transcend to the entire supply chain. Once a determination has been made regarding the usefulness of RFID in one firm, an analysis should then be extended across the supply chain to determine if performance holds for all supply chain members.

# CHAPTER 2
# RFID 101

RFID is a technology that uses radio frequencies to identify, track, and trace an object or product. As with many modern technologies, RFID technology has its origin in military applications during World War II with commercial applications beginning during the early 1980s. Today, these applications span several industries and are not limited to the Wal-Mart and associated mandates within the retail industry that have sparked recent interest about RFID. In fact RFID has, for the most part, been flying below the business innovation radar.

This chapter begins with a history of RFID followed by an overview of RFID system components. The technology's advantages and limitations are discussed in the final sections of the chapter.

## RFID History

### Early History of RFID and Overview to Present

RFID was originally used by the military to identify friend or foe aircraft (IFF) during the Second World War. By no means can we say that RFID is a new technology, although it took a few decades to leave the realm of scientific research and become practical for business applications. It is difficult to trace its true history because most research was done behind closed doors for military purposes. However, RFID uses electromagnetic energy and its roots can be traced to research on the use of radar. One of the first major papers on RFID was by Harry Stockman: "Communication by means of reflected power," published in October 1948, which described the technology as imperfect and said that much research and development would be needed before RFID technology could be put into practical use in various fields. Other important early papers on RFID include:

- Application of the microwave homodyne (Vernon, 1952*)

---

**Source*: Vernon, F. (1952) "Application of the microwave homodyne," *IRE Transactions on Antennas and Propagation*, AP-4: 110–116.

- Radio transmission systems with modulatable passive responder (Harris, 1960[†])
- Electronic identification system (Klensch, 1975[‡])

Since then, the number of research studies and the number of patents filed have continued to increase. By 2000, over 350 patents related to RFID and its use had been filed. Commercial applications began to flourish in the past few decades. Early companies such as Checkpoint, Sensormatic, and Togo developed electronic article surveillance (EAS) systems to prevent retail store theft, primarily for higher-priced items. This commercial use of RFID was quite effective in preventing theft. Major events in RFID commercial development can be traced to the 1975 declassification of research by Los Alamos Scientific Labs (LASL) with the published paper "Short-range radio-telemetry for electronic identification using modulated backscatter" (authored by A. R. Koelle et al.).

By the 1980s, interest in RFID began to grow rapidly. The initial development efforts in the United States focused primarily on transportation applications and personnel access. In Europe, the efforts were focused on short-range systems for animals followed by industrial and business applications. Although testing of RFID for collecting tolls had been conducted for many years in the United States, the first commercial toll application began in Europe in 1987, which was soon followed in the United States. During the 1990s, RFID technology continued to be implemented around the world. The first open-highway electronic tolling system, where vehicles could pass toll collection points at highway speeds, was instituted in Oklahoma in 1991. Then followed the first combined toll collection and traffic management system, installed in Houston, Texas, in 1992. Other RFID toll-tag developments such as multiple uses of the same tag for different purposes followed. For example, one tag could be used for both toll collection and access control (parking lot, gated community, etc.).

The twenty-first century has seen an explosion in the use of RFID technology for supply chain management applications (Fig. 2.1). Many companies are now using RFID tags to track individual inventory items, cases, pallets, or equipment. RFID tags are being used within manufacturing plants, warehouses, and retail facilities. Some companies are also tagging individual parts that go into the finished product.

## History of Later RFID Developments

Commercially, the technology has been applied since the 1980s, with increased acceptance by the mid-1990s. Case uses include:

- Keyless entry and smart tickets (Fig. 2.2)
- Document information and smart stamps (Fig. 2.3)

---

[†]*Source*: Harris, D.B. (1960) "Radio transmission systems with modulatable passive responder," *US Patent* 2,927,321.

[‡]*Source*: Klensch, R. (1975) "Electronic identification system," *US Patent* 3,914,762.

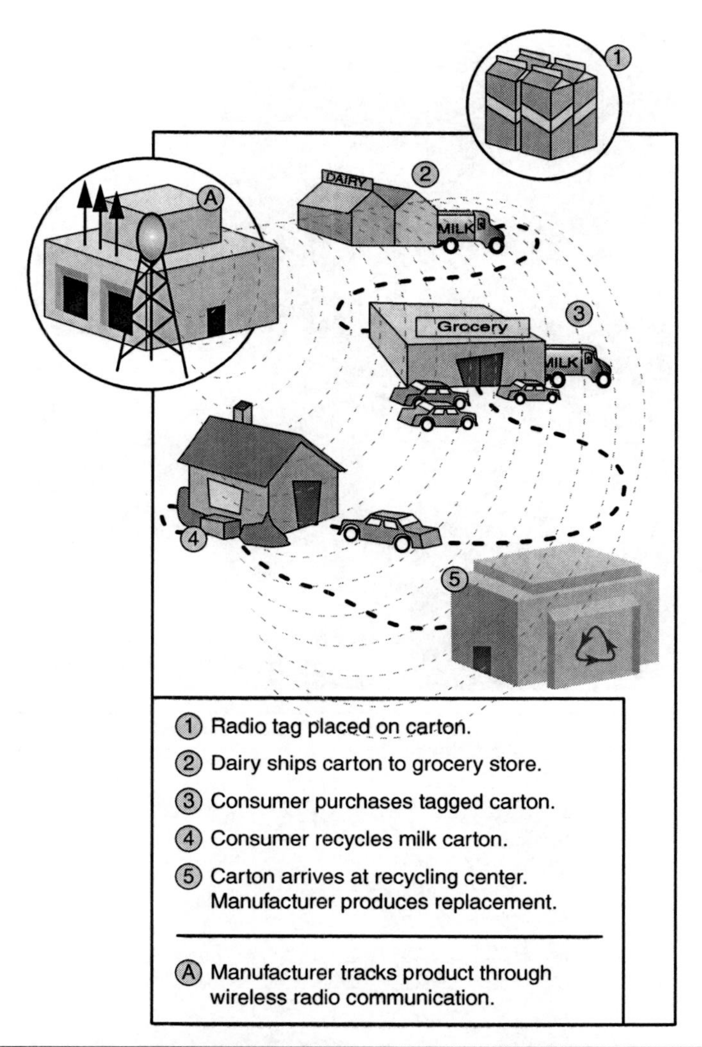


**Figure 2.1** RFID for supply chain applications. (*Source*: Courtesy of HowStuffWorks.com)

- Badge readers (Fig. 2.4)
- Automatic highway and bridge toll collection (Fig. 2.5)
- Offender tags
- Tracing livestock movements (Fig. 2.6)
- Tracking and control of nuclear inventories
- Tracking air freight
- Automobile manufacturing through assembly lines
- Health care

**Figure 2.2** Keyless entry. (*Source*: http://media.marketwire.com/attachments/200910/TN572998_SmartBandopeningSmarteLockeLocker.jpg)

**Figure 2.3** Document management. (*Source*: http://www.virtualdoxx.com/images/color_coded_files10.jpg)

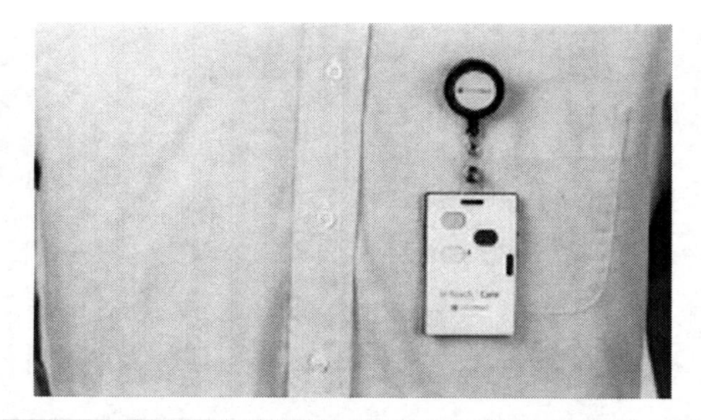

**FIGURE 2.4** RFID Badge. (*Source*: http://d2eosjbgw49cu5.cloudfront.net/rfid-weblog.com/imgname--it740_staff_badge_worldas_thinnest_hybrid_active_rfid_tag--50226711-images--IT740-RFID-Staff-Badge.jpg )

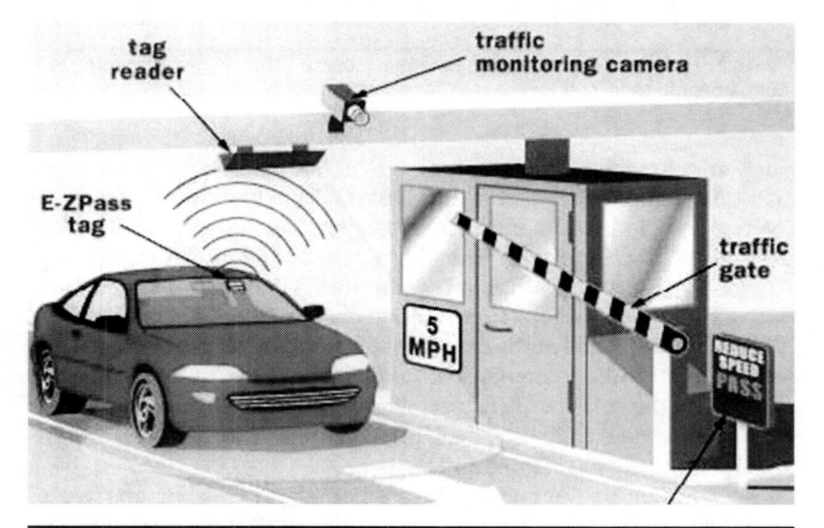


**FIGURE 2.5** The Port Authority of New York and New Jersey pioneered the idea of using RFID in highway toll booth collection. (*Source*: http://www.witiger.com/ecommerce/300px-SR_417_University_Toll_Plaza.jpg)

- Tracking railroad assets
- Tracking military asset
- Law enforcement
- Libraries

## Research on RFID in Supply Chains

Compliance mandates by the U.S. Department of Defense and Wal-Mart requiring suppliers to use Electronic Product Code (EPC) RFID tagging on pallets and cases (January 2005) have provided a strong

**Figure 2.6** Tracing livestock movements. (*Source*: http://www.lancasterdhia.com/images/tags_calf.jpg)

push to mainstream the adoption of RFID in the supply chain, especially by manufacturers and distributors. This big bang sparked an interest that led to other RFID initiatives and research. Other key players driving the growth and adoption include Tesco, Metro Group, Albertson's, and Target. Table 2.1 summarizes some of the key mandated initiatives. Tesco, a supermarket chain in the United Kingdom, has introduced RFID at the case level for tracking shipments between its 2000 distribution centers and retail stores. The Metro Group has focused on the store of the future, while Albertson's (the second largest supermarket chain in the United States) also has RFID pilot-studies in the Dallas, Texas, area.

In addition to the retail industry, Boeing and Airbus are two of many other companies that have implemented RFID to track inventories (including work in progress). Michelin has tested RFID transponders embedded into tires, where the primary purpose is to make tire-tracking easier to comply with the U.S. Transportation, Recall, Enhancement, Accountability, and Documentation Act (TREAD Act).

## RFID Today

Before 2000, most of the commercial RFID implementations were associated with animal tracking and access control (particularly transportation and retail theft prevention). Among the earliest uses of RFID was tracking livestock movement, mainly for inventory management. This was followed by RFID access control.

In its simplest form, RFID access control involves allowing certain people access to an area or resource while barring others from a

| Company | Objective | Target Implementation Date |
|---|---|---|
| Michelin North America Inc. | Implanted RFID tags on selected tires to keep track of their performance and tread wear over a period of time. | January 2003 |
| Metro Group (Europe) | Pilot-testing RFID in the supply chain and warehouse of the Extra Future Store in Germany. | April 2003 |
| Tesco Corp. (UK) | RFID tags are placed on cases of nonfood items at the retailer's distribution centers and tracked into stores. | April 2004 |
| | Unspecified number of suppliers are required to tag cases delivered to some Tesco distribution centers. | September 2004 |
| Target | Unspecified number of top suppliers are required to apply tags to pallets and cases to selected regional distribution centers. | Late spring 2005 |
| Boeing Co. | Suppliers are asked to place RFID tags on parts so they can be traced through their life cycle for manufacturing and maintenance. | To be determined |

*Source*: Reyes, P.M. and Frazier, G.V. (2007). "Radio frequency identification: past, present, and future business applications," *International Journal of Integrated Supply Management*, 3(2): 125–134.

**Table 2.1** RFID Initiatives

similar level of access. The systems used to restrict entry can prevent individuals from looking at sensitive data, taking advantage of various resources, gaining entry into areas for which they lack authorization, and even leaving such areas. The two primary applications of RFID access control began with transportation and retail theft prevention. In transportation access control, vehicles gained access to highways or bridges via toll tag. Retail theft prevention, used mainly for high-priced retail items, caused an alarm to sound if an item left the store without having the RFID tag removed by the cashier.

Recent RFID applications are more diverse, ranging from cashless payment at gasoline stations to monitoring of paroled criminals. Additional current examples of RFID applications include Hewlett-Packard placing RFID tags on its printer boxes and Delta Air Lines testing the use of RFID tags for tracking luggage. RFID tags on books are being used in many university library systems (including the University of Michigan and the University of Texas at Arlington) to facilitate students checking books in or out. JPMorgan Chase, a large U.S. financial services company, has started using RFID chips embedded in its credit cards (nicknamed "blink"), which allows the consumer to hold the card close to a reader instead of swiping the card through a reader and risking the magnetic strip being damaged.

Other examples of applications can frequently be found in current trade publications like *IDTechEx* (www.idtechex.com) and *RFID Journal* (www.rfidjournal.com):

- Japanese students' uniforms were tagged with RFID to track their movements. Students at Brandon Junior High School in the United Kingdom wore RFID tags on their uniforms and were counted and identified during a mass emergency evacuation.
- Jiahgsu Logton Jail in China enhanced its security procedures by using RFID-based prisoner identification.
- Boeing and FedEx jointly tested RFID tags on in-service aircraft parts to improve maintenance and reliability.
- Merck, Novartis, and other U.K. pharmaceutical companies ran pilot tests to detect counterfeit drugs and dispensing errors before the drugs reached patients.

One factor that has further facilitated RFID supply chain applications is the adoption of a global standard for data formats in tags. This Electronic Product Code (EPC) standard was developed by the Auto-ID Center, a consortium founded in 1999 by five leading research universities, anchored by the Massachusetts Institute of Technology and nearly 100 leading retailers, consumer products manufacturers, and software companies. In the mid-2000s the world's largest retailer, Wal-Mart, created a surge in RFID implementations by mandating that its largest 300 suppliers start using RFID tags on cases shipped to Wal-Mart warehouses. Wal-Mart's desired outcome with RFID is to improve its overall supply chain operations by increasing inventory visibility, reducing theft, and reducing the overall cost of logistical operations while keeping track of inventory movements. Over the long term, some believe that companies will have little choice but to adopt RFID to remain competitive.

## RFID System Components

In general, an RFID system consists of tags, antennas, readers, and communication infrastructure (Fig. 2.7). Typically, an RFID transponder or tag is attached to an object. The RFID transponder or tag can be active (battery powered and proactively emitting a radio frequency signal) or passive (unpowered and reactively emitting a radio frequency signal). An RFID reader communicates with the transponder or tag to identify the object to which the transponder or tag is attached. The transponder or tag has information about the object for identification purposes depending upon the need. Stored within a transponder or tag could be the serial number, model number, or other characteristics of the object to identify or distinguish this object from others or to track the movement of the object.

In this section, we focus our attention on the main RFID system components that pertain to commonly observed RFID systems. The following subsections provide the reader with a brief overview of the RFID system components.

### RFID Tags

An RFID tag (also called a transponder) is basically a microprocessor chip consisting of an integrated circuit with memory (Fig. 2.8). In general, the RFID tags can be grouped into basic categories: type, frequencies, or by capabilities.

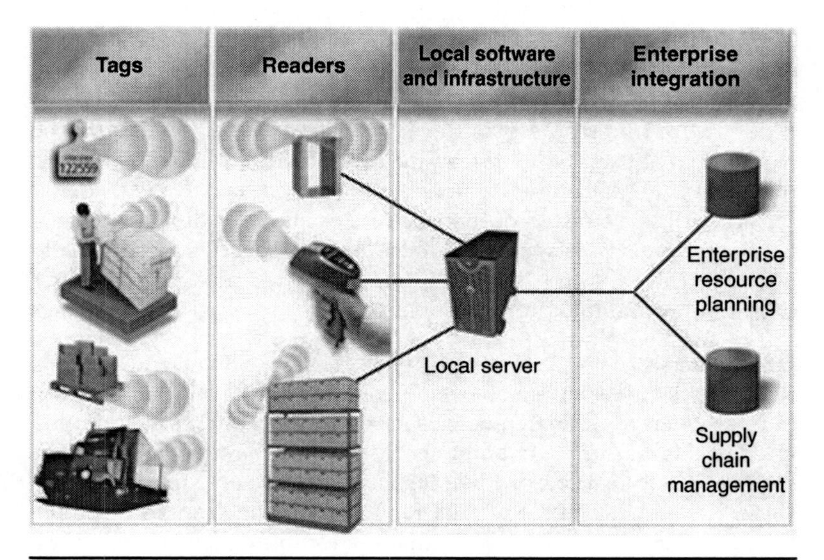


**Figure 2.7** RFID system components. (*Source*: http://www.digitivity.com/articles/picture_rfid_technology.jpg)

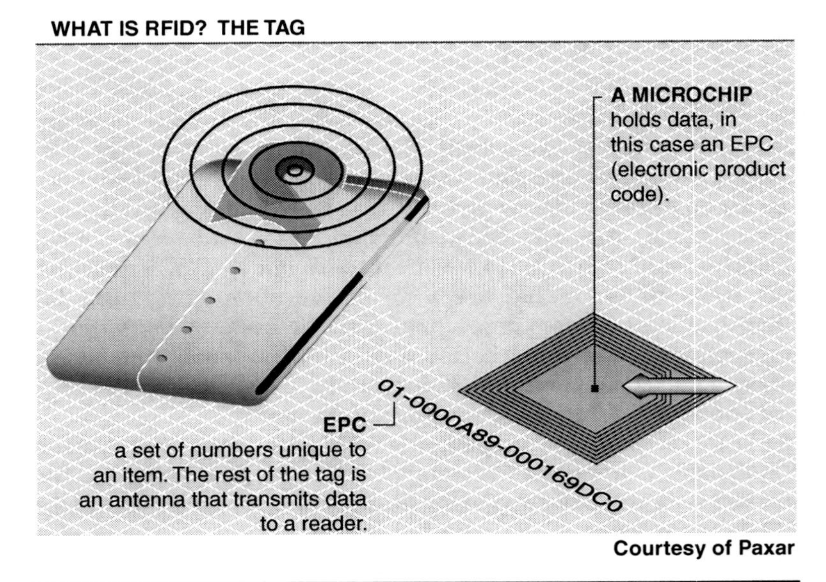


**FIGURE 2.8** The RFID tag. (*Source*: http://www.tx.ncsu.edu/jtatm/volume4issue3/images/RFID.jpg)

## Types

The RFID tags can be grouped into basic types—passive and active—which differ based primarily on whether the tag contains some form of onboard power supply (i.e., a battery). The primary function of the tag is to transmit data to the RFID system. The differences between passive and active tags are summarized in Table 2.2. The passive tags (Fig. 2.9), which are generally less expensive than the active tags, do not contain an onboard power supply. Instead, passive tags use power produced by the reader in order to transmit the stored data to the reader. Active tags (Fig. 2.10), on the other hand, contain an onboard power supply that is used to transmit the stored data to the reader.

A third type consists of semiactive tags and is designed with features of both passive and active tags. These semiactive tags typically use an internal battery, and attempts to use the advantage of each type, while eliminating the disadvantages.

## Frequencies

RFID applications have also been developed across a broad range of radio frequencies. The frequencies used in RFID systems are grouped into low, high, ultra high, and microwave. The characteristics of each range and applications of RFID tags in the range are summarized in Table 2.3, and examples of tags for the frequency ranges are illustrated in Figs. 2.11 to 2.13. Note that the radio frequency range for a tag relates strongly to its read range and per unit cost, making a tag more or less useful for certain operational applications.

| Passive Tags | Active Tags |
|---|---|
| Operate without a battery | Powered by an internal battery |
| Less expensive | More expensive |
| Unlimited life (because of no battery) | Finite lifetime (because of battery) |
| Less weight (because of no battery) | Greater weight (because of battery) |
| Lesser range (up to 3–5 m, or 10–16 ft) | Greater range (up to 100 m, or 330 ft) |
| Subject to noise | Better noise immunity |
| Gets power from reader | Internal power to transmit signal |
| Requires more powerful readers | Can be effective with less powerful readers |
| Lower data transmission rates | Higher data transmission rates |
| Fewer tags can be read simultaneously | More tags can be read simultaneously |
| Greater orientation sensitivity | Less orientation sensitivity |

*Source*: Adopted from Wyld, D. (2006), "RFID 101: the next big thing for management," *Management Research News*, 29 (4): 154–173.

**TABLE 2.2** Differences in Passive and Active RFID Tags

**FIGURE 2.9** Passive tag. (Shown here is an Alien Squiggles tag.) (*Source*: http://www.alientechnology.com/docs/products/DS_ALN_9640.pdf)

**FIGURE 2.10** HF tag. (*Source*: http://www.sagedata.com/images/2007/RFID%20Tag%20HF.jpg)

| Frequency Band | System Characteristics | Example Applications |
|---|---|---|
| Low (LF) 100–500 KHz (typically 125–134 KHz worldwide) | Short read range (to 0.5 m [18 in.])<br>Low reading speed<br>Relatively inexpensive<br>Can read through liquids<br>Works well near metals | Access control<br>Animal identification<br>Beer keg tracking<br>Automobile key/ antitheft systems |
| High (HF) (typically 13.56 MHz) (refer to Figs. 2.11 and 2.13 for examples) | Short to medium read rates (1–3 m [3–10 ft])<br>Medium reading speed<br>Can read through liquids; works well in moist environments<br>Does not work well near metal<br>Moderately expensive | Access control<br>Smart cards<br>Electronic article surveillance<br>Library book tracking<br>Pallet/container tracking<br>Airline baggage tracking<br>Apparel/laundry item tracking |
| Ultra high (UHF) 400–1000 MHz (typically 850–950 MHz) (refer to Fig. 2.12 for example) | Long read range (3–9 m [10–30 ft])<br>High reading speed<br>Reduce likelihood of signal collision<br>Difficulty reading through liquids<br>Does not work well in moist environment<br>Subject to interference from metals<br>Relatively expensive | Item management<br>Supply chain management |
| Microwave 2.4–6.0 GHz (typically 2.45 or 5.8 GHz) | Medium read range (3+ m [10+ ft])<br>Similar characteristics to UHF tags, but with faster read rates | Railroad car monitoring<br>Toll collection systems |

*Source*: Adpted from Wyld, D. (2006), "RFID 101: The next big thing for management," *Management Research News*, 29(4): 154–173.

**TABLE 2.3** System Characteristics and Applications of RFID Tags

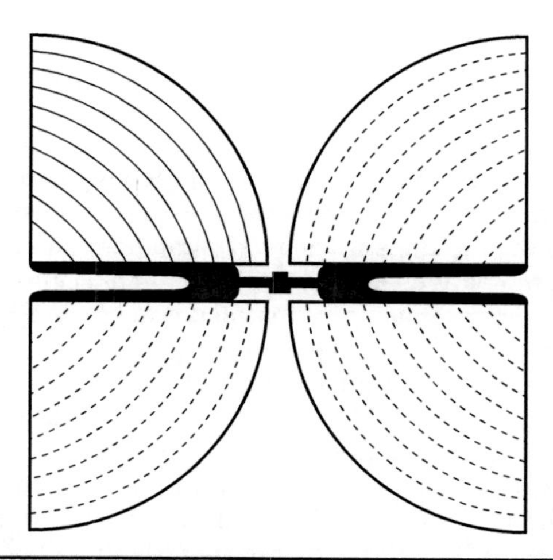

**Figure 2.11** RFID tag used for asset tracking. (*Source*: http://www.gaorfidassettracking.com/RFID_Asset_Tracking_Products/images/rfid_tag_865_115002.gif)

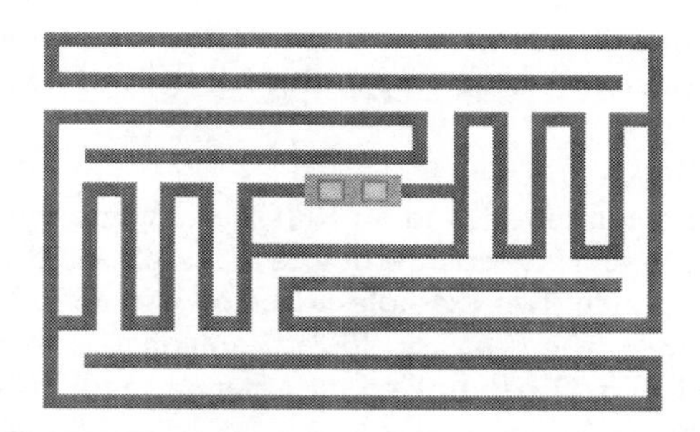

**Figure 2.12** EPC RFID tag. (*Source*: http://foodconsumer.org/7777/uploads/1/EPC-RFID-TAG.jpg)

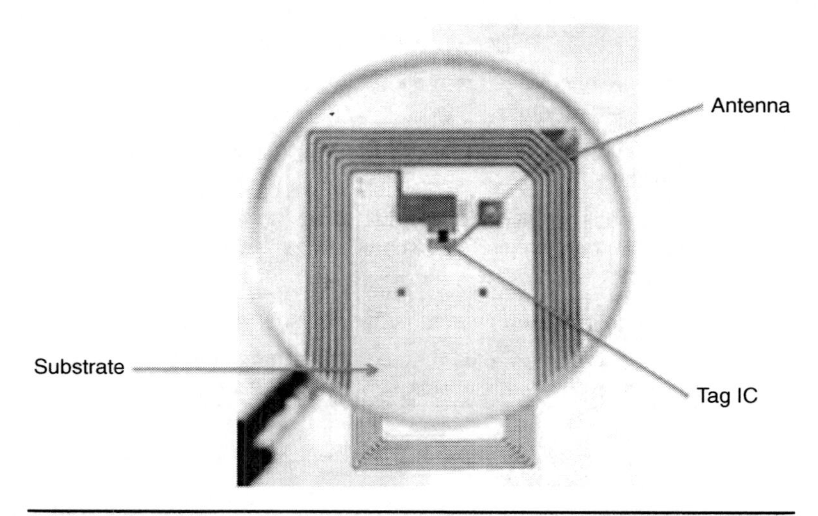


**Figure 2.13** RFID passive tag. (*Source*: http://rob.sh/img/rfid_passive_tag.png)

### Capabilities

RFID tags can also be categorized by their memory capabilities: read-only, read/write tags, or combination tags. The read-only tags store data that cannot be changed. The read/write tags store data that can be altered. Combination tags have some data that cannot be changed along with memory capacity for updated data. Most RFID tags have limited memory capacity, as they are designed not as a repository of comprehensive object/product data, but rather as a pointer to an Electronic Product Code (EPC) system or other database that contains the complete information stored for that item.

Data storage capabilities of RFID tags for product data have been defined by the EPC framework, an RFID industry standard developed by the Auto-ID Center for product identification by the consumer packaged goods industry. The EPC framework consists of six different classes of tags, with each class having an ascending range of capabilities. These classes are outlined in Table 2.4. The EPC framework is only one of several such frameworks, as different industries require different data storage capabilities of their RFID tags.

### RFID Antenna

The RFID tag communicates to an RFID reader through the reader's antenna. The antenna is a separate device that is physically connected to the reader by a cable (two examples are shown in Figs. 2.14 and 2.15). Typically, a single reader has up to four antennas with cable lengths between 6 and 25 ft. Depending on the antenna's radio wave footprint,

| EPC Tag Class | Tag Class Capabilities |
|---|---|
| 0 | EPC number is factory programmed onto the tag and is read-only |
| 1 | Read/write-once tags are manufactured without EPC number (user programmable) |
| 2 | Class 1, plus larger memory and encryption and read/write capabilities |
| 3 | Class 2 capabilities, plus a power source to provide increased range and/or advanced functionality (such as sensing capability) |
| 4 | Class 3 capabilities, plus an active transmitter and sensing |
| 5 | Class 4 capabilities, plus the ability to communicate with passive tags (essentially acts as a reader) |

*Source*: Adopted from Wyld, D. (2006), "RFID 101: The next big thing for management," *Management Research News*, 29(4): 154–173.

**Table 2.4** EPC Tag Classes

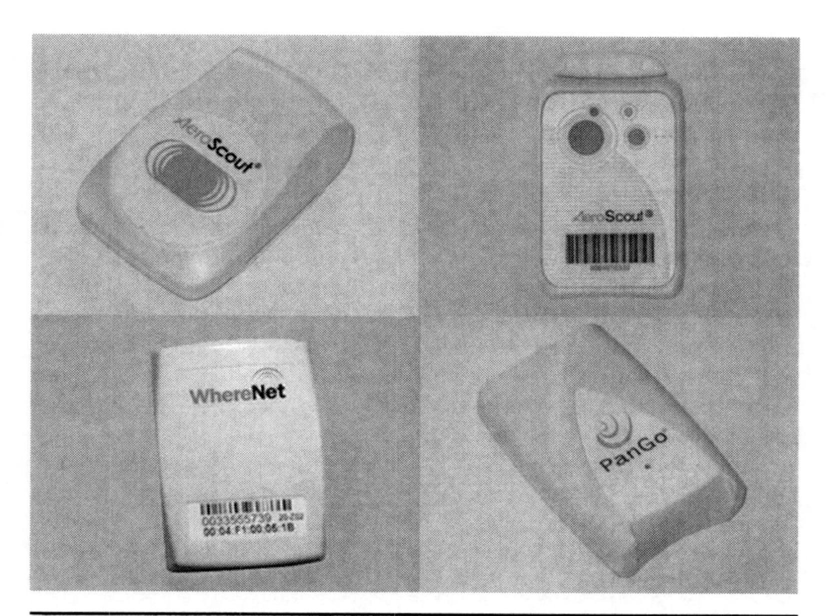


**FIGURE 2.14** RFID active tags. (*Source*: web-help@cisco.com)

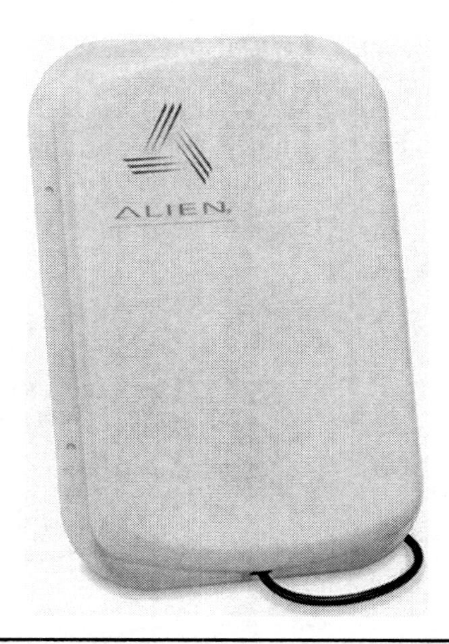


**FIGURE 2.15** Alien Antenna. (*Source*: Sales@alientechnology.com)

an RFID antenna can have a reading range in front of the antenna as well as sideways. The antenna broadcasts the reader's RF signal and receives the tag response. Antennas need to be properly positioned relative to tags in order to obtain good tag reading accuracy.

## RFID Multiplexers

In some RFID applications, there are compelling reasons for using one reader and multiplexing (or switching) through multiple antennas. This is particularly true with the need to read tags in multiple locations, or zones, without the associated cost of multiple readers and multiple host controllers. Another example would be with a low-power transceiver (transmitting/receiving reader) in a system design calling for complete coverage in an area where a magnetic "hole" would exist if one larger antenna was used with medium to small tags.

The multiplexing concept and method attempt to assist designers and users of embedded low-power HF RFID systems (for example) achieve cost targets without sacrificing system performance. In general, a multiplexed RFID reader system will consist of one reader and multiple antennas.

## RFID Readers

An RFID reader (Fig. 2.16), also known as an interrogator, is a device that can read and/or write data to an RFID tag. The communication between the reader's antenna and nearby RFID tags facilitates the transfer of needed information about an RFID-tagged object. Because the tags operate on certain frequencies, so must the readers. For communication compatibility, the RFID tags must operate at the same radio frequency and use the same protocols as the RFID reader. Readers come in many form factors (i.e., they come with varying physical characteristics), including handheld, vehicle mounted, post mounted,

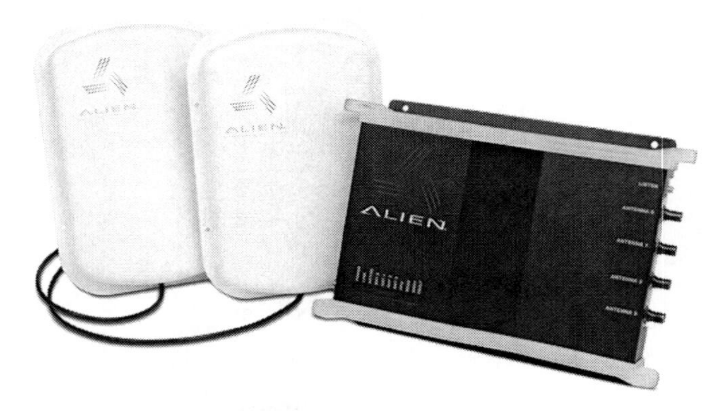


**FIGURE 2.16** RFID readers. (*Source*: http://www.gaorfid.com/images/RFID-Readers-Gen2(UHF)/RFID-Readers-Gen2EPC(UHF)4.bmp)

and hybrid. Dedicated readers specifically read either passive or active tags, and hybrid readers can switch between reading the two types of tags. Alien, for example, is a multiprotocol reader—reading multiple types of tags (active or passive, Gen1 or Gen2).

## RFID Printers

An RFID reader is a type of stationary reader that can print a bar code and also write an RFID tag—together referred to as a smart label. This smart label consists of a bar code with an embedded RFID tag. Many types of information can be printed on the label. As the smart label is printed, the RFID printer reads the smart label to validate what was written as part of the encoding process. Examples of RFID printers are provided in Fig. 2.17.

## Communication Infrastructure

As with any digital data, the data generated by RFID systems can be transferred across supporting communication infrastructure. This infrastructure includes the physical network communication technology, wired and/or wireless, that carries and stores the data. It provides connectivity and security for the different components of an RFID system (Fig. 2.18).

**Figure 2.17** A Sato RFID printer. (*Source*: http://www.satothermalprinters.com/images/printers/CLe-rfid.jpg)

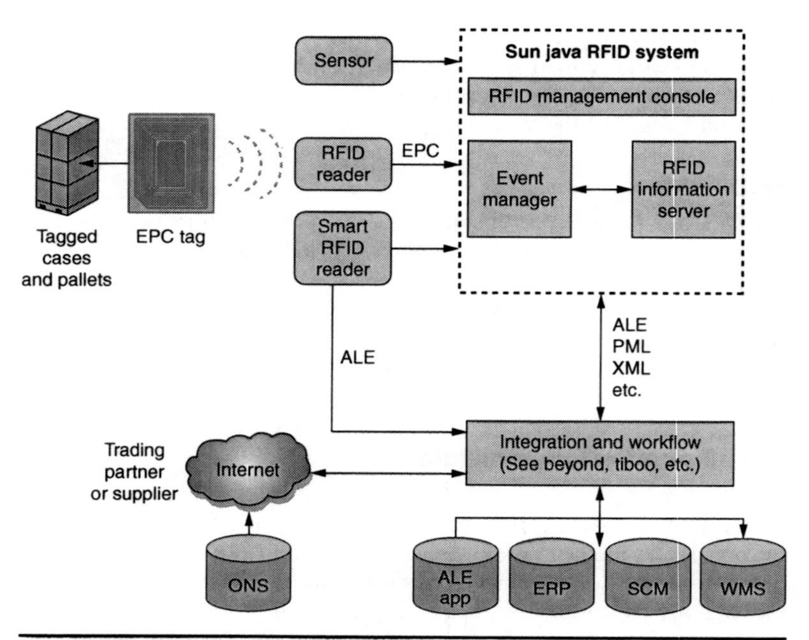


**FIGURE 2.18** RFID communication infrastructure. (*Source*: http://docs.sun.com/source/819-4684/figures/RFID-intro-4.gif)

In addition to the basic system components, an RFID system may also include a controller, sensor, annunciator, actuator, and a host and software system. The controller, similar to driver software, is an intermediary agent that provides a communication interface allowing an external entity to communicate (and control) a reader's behavior. A sensor, attached to a reader, can be used to turn the reader on or off based on some event detected by the sensor. An annunciator is an electronic signal or indicator, such as alarms or light stacks, used to alert users to various statuses of different system attributes. An actuator is a mechanical device for controlling or moving objects.

The host and software system is a general term used to describe the hardware and software components that are separate from the basic RFID system and is a compilation of an edge interface system, any middleware, an enterprise back-end interface, and the enterprise back-end. The edge interface system component integrates the host and software system with the basic RFID system. This component's primary task is to get data from the reader and control the readers' behavior. This component can perform many tasks, such as filtering duplicate reads from different readers, and can be used to provide intelligent functions. The middleware has been broadly defined as everything that is between the edge interface and the enterprise back-end interface. Acting similar to the human body's nervous system, this component provides core functionality of the system. The enterprise back-end interface connects the middleware with the enterprise back-end component and is typically where the business process is integrated with the basic RFID system. The enterprise back-end component, in general, is the data repository, provides the data directory for the tagged objects, and is the overall process engine for the enterprise.

## Advantages of RFID Technology

Until recently, bar codes were the most prevalent technology for object or product identification. Bar codes and RFID each have their own unique physical properties that make them easy or difficult to read under certain environmental conditions. Bar codes require an uninterrupted line of sight to be visible to a bar code reader, while RFID tags can be packaged either inside or outside an object and still be read. Bar code reading is impaired by dirt, moisture, abrasion, and packaging contour, while RFID is not as susceptible to those conditions. However, RFID tags can be affected by metal and liquid properties of a product to which they are affixed. RFID tags may have read/write capability, whereas bar codes do not have such capabilities; therefore more data can be stored on an RFID tag than on a bar code.

RFID provides a means to automatically identify and track items using tags that provide information in real time about their identity, location, activity, or history, which is then processed and utilized by application software. RFID systems for the supply chain emphasize tagging of pallets, cases, and (in certain situations) individual items. In contrast to bar codes, which are used by more than a million firms in over 140 countries and 23 industries, RFID employs radio frequencies to transmit data to readers within a certain distance. RFID also offers several key technological advantages over bar codes, which are summarized in Table 2.5.

A key benefit of RFID over bar codes is made possible by the Internet and its underlying information infrastructure. The richness and timely availability of information about the location and status of goods worldwide to manufacturers, distributors, and retailers is what motivated retailers, such as Wal-Mart, and the U.S. Department of Defense to mandate the use of RFID by top suppliers.

*Absence of line of sight.* A line of sight is not required for an RFID reader to read an RFID tag. This is perhaps the most compelling advantage of RFID. An RFID reader can read a tag through obstructing material that is RF-lucent for the frequency used.

*Contactless.* An RFID tag can be read without any physical contact between the tag and the reader. Among the advantages are absence of wear and tear on the tags for reading and writing, as well as the

| Bar Codes | RFID Tags |
|---|---|
| Bar codes require line of sight to be read | RFID tags can be read or updated without line of sight |
| Bar codes can only be read individually | Multiple RFID tags can be read simultaneously and with greater speed and efficiency |
| Bar codes must be visible to be logged | RFID tags can be read even when concealed within an item |
| Bar codes cannot be read if they become dirty or damaged | RFID tags are able to cope with harsh and dirty environments |
| Bar codes are fixed at the time of printing | Rewritable RFID tags can be reused for multiple applications, lowering cost of ownership |
| Bar codes must be manually tracked for item identification, making human error an issue | RFID tags allow complete automated data handling for paper reduction and greater overall efficiency, and automatic tracking, eliminating human error |

**TABLE 2.5** Key Technological Advantages of RFID over Bar Codes

readers. The biggest advantage is that operations are not slowed down in order for the reader to physically contact the tag.

*Support for multiple tag reads.* Support for multiple tag reads is another important advantage of RFID. By using what is called an anticollision algorithm, it is possible to use an RFID reader to automatically read several RFID tags in its read zone within a short period of time. Depending on the RFID tag and business application, this advantage allows a reader to uniquely identify from a few up to many tags per second. Hence, the data collected from the tagged objects—whether moving or stationary—is a technology advantage compared to having to read one tag at a time.

*Rugged.* Passive RFID tags can sustain rough operational environment conditions, such as heat, humidity, cold, corrosive chemicals, and mechanical vibrations to a fair extent. Some passive tags can survive temperatures ranging from –40°C to 200°C (–40°F to 400°F). In general, these tags are made to handle a specific application and operating environment.

*Writable data.* The data of a read/write (RW) RFID tag can be rewritten up to 100,000 (or more) times. However, the more common tag currently used is a write once, read many (WORM) tag.

*Variety of read ranges.* An RFID tag can have a read range from a few inches to more than 100 ft, depending on the frequency of the tag. A low-frequency (LF) passive tag has a read distance of just a few inches, while a high-frequency (HF) passive tag has a read distance of 3 ft. An ultra-high frequency (UHF) passive tag can have a read distance of 300 ft; and an active tag in the gigahertz range can have a read distance greater than 100 ft.

*Wide data capacity range.* A passive RFID tag can store from a few bytes of data to hundreds of bytes. Active RFID tags can store virtually any amount of data and are not limited in their capacity range because the physical dimensions and capabilities of active tags are not limited.

*Smart tasks.* In addition to being a carrier and transmitter of data, an RFID tag can be designed to perform other duties, such as measuring temperature.

## Limitations of RFID Technology

Like other technologies, RFID has its limitations. In fact, some of the advantages have limitations.

*Performance.* An RFID reader could partially or completely fail to read the tag data as a result of RF-opaque material, RF-absorbent material, or frequency interference.

*Environmental factors.* Depending on the frequency, the read accuracy of the tags could be affected if the operations environment has large amounts of metals and liquids.

*Actual tag reads.* Because the reader has to use some kind of anti-collision algorithm, the number of actual tags that a reader can uniquely identify (per unit of time) is limited.

*Hardware interference.* If the RFID readers are improperly installed, then it is possible for the readers to show evidence of reader collision.

*Penetrating power of the RF energy.* The penetrating power of the RF energy is dependent on the reader's transmitting power and its duty cycle. For example, if cases on a pallet are stacked too deep, then it is possible that a reader may fail to read some of the cases.

*Immature technology.* While RFID technology has been around for many decades, the sparked interest has also sparked interest in types of applications. An RFID solution may not be readily available, and thus vendors need to develop the products. The issue of maturity will continue.

# CHAPTER 3

# EPCglobal Overview and Standards*

EPCglobal is leading the development of industry-driven standards for the Electronic Product Code (EPC) to support the use of radio frequency identification in today's fast-moving, information-rich trading networks. It is a subscriber-driven organization comprising industry leaders and organizations focused on creating global standards for the EPCglobal Network. Its goal is to increase visibility and efficiency throughout the supply chain and enable higher-quality information flow between companies and their key trading partners.

EPCglobal provides the following services to companies wishing to improve their supply chain management efficiency:

- Assignment, maintenance, and registration of EPC Manager Numbers
- Participation in development of EPCglobal Standards via EPCglobal's Action and Working Groups
- Access to the EPCglobal Standards, research, and specifications
- Opportunity to influence the future direction of research by the Auto-ID Labs
- Access to the results of the EPCglobal Certification and Accreditation Program testing
- Links with other subscribers to create pilots and test cases
- Training and education on implementing and using EPC technology and the EPCglobal Network through more than 101 global agents

EPCglobal is one of the standards organizations under the umbrella of GS1, which also includes subscriber organizations working on standards for bar codes, electronic data interchange (EDI), and a global data synchronization network (GDSN).

---

**Source*: www.epcglobalinc.org

## EPC Tags

The structure of EPC tags was first developed at the Auto-ID Center at the Massachusetts Institute of Technology, and is now managed by EPCglobal (www.epcglobalinc.org), a not-for-profit joint venture between EAN International and the Uniform Code Council. This body manages UPC (Universal Product Code) information in bar codes and sets the standards for how basic product information is encoded in RFID chips and how information is passed from RFID readers to various applications, as well as from application to application. EPC represents a specific approach to item identification, including an emerging standard for the tags. The current version of the EPC Tag Data Standard (see Fig. 3.1) specifies the data format for encoding and reading data from 64- and 96-bit RFID tags. The EPC-compliant tag structure consists of a number made up of a header and three sets of data. The header identifies the EPC's version number. The second part of the number identifies the EPC Manager (usually the manufacturer of the product to which the EPC is attached). The third, called object class, refers to the exact type of product, most often the SKU. The fourth, the serial number, is unique to the item, telling specifically what is being identified and making it possible to rapidly find products nearing their expiration date or to manage product recalls.

The following is a description of tag class definitions (last updated by EPCglobal November 1, 2007).

### Class-1: Identity Tags

Passive-backscatter tags with the following minimum features:

- An Electronic Product Code (EPC) identifier
- A tag identifier (tag ID)

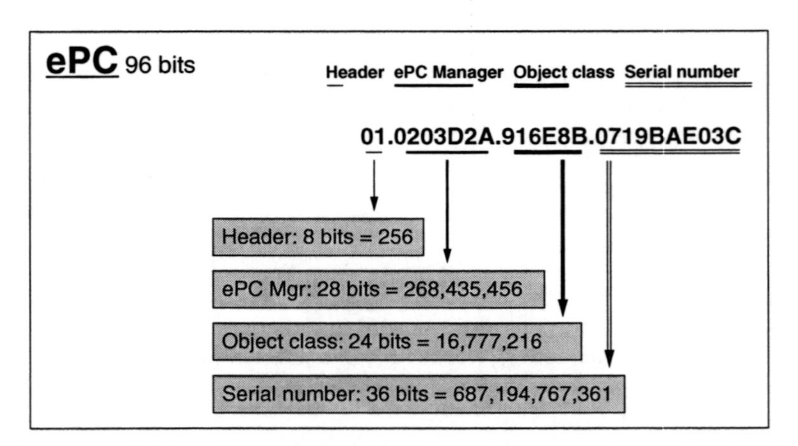


**Figure 3.1** EPC tag structure. (*Source*: http://www.epcglobalinc.org/standards)

- A function that renders a tag permanently nonresponsive
- Optional decommissioning or recommissioning of the tag
- Optional password-protected access control
- Optional user memory

### Class-2: Higher-Functionality Tags

Passive tags with the following anticipated features above and beyond those of Class-1 tags:

- An extended tag ID
- Extended user memory
- Authenticated access control
- Additional features (TBD) as defined in the Class-2 specification

### Class-3: Battery-Assisted Passive Tags (called Semipassive Tags in UHF Gen 2)

Passive tags with either or both of the following anticipated features above and beyond those of Class-2 tags:

- A power source that may supply power to the tag and/or to its sensors
- Sensors with optional data logging

Class-3 tags still communicate passively, meaning that they (1) require an "interrogator" (reader) to initiate communications, and (2) send information to an interrogator using either backscatter or load-modulation techniques.

### Class-4: Active Tags

Active tags with all of the following anticipated features:

- An Electronic Product Code (EPC) identifier
- An extended Tag ID
- Authenticated access control
- A power source
- Communications via an autonomous transmitter
- Optional user memory
- Optional sensors with or without data logging

Class-4 tags have access to a transmitter and can typically initiate communications with an interrogator or with another tag. Protocols may limit this ability by requiring an interrogator to initiate or enable tag communications. Because the active tags have access to a transmitter, of necessity they have access to a power source. Class-4 tags shall not interfere with the communications protocols used by Class-1/2/3 tags.

## Standards

Standards and specifications provide the common definitions, functionality, and language for the hardware and software components of the EPCglobal Network. They help advance the EPCglobal community toward a common objective, namely, implementing the EPCglobal Network to improve visibility and efficiency in today's global, multi-industry supply chain.

## EPCglobal Architecture Framework

This standard defines and describes the EPCglobal Architecture Framework (Fig. 3.2). The EPCglobal Architecture Framework is a collection of interrelated standards for hardware, software, and data interfaces, together with core services that are operated by EPCglobal and its delegates, all in service of a common goal of enhancing the supply chain through the use of Electronic Product Codes (EPCs). This document has several aims:

- To enumerate, at a high level, each of the hardware, software, and data standards that are part of the EPCglobal Architecture Framework and show how they are related
- To define the top-level architecture of core services that are operated by EPCglobal and its delegates
- To explain the underlying principles that have guided the design of individual standards and core service components within the EPCglobal Architecture Framework
- To provide architectural guidance to end users and technology vendors seeking to implement EPCglobal standards and to use EPCglobal core services

The next three sections give an overview of standards at the three architectural levels: exchange, capture, and identify (Fig. 3.2).

### Exchange

#### EPCglobal Certificate Profile Standard

To ensure broad interoperability and rapid deployment while ensuring secure usage, this document defines a profile of X.509 certificate issuance and usage by entities in the EPCglobal network. The profiles

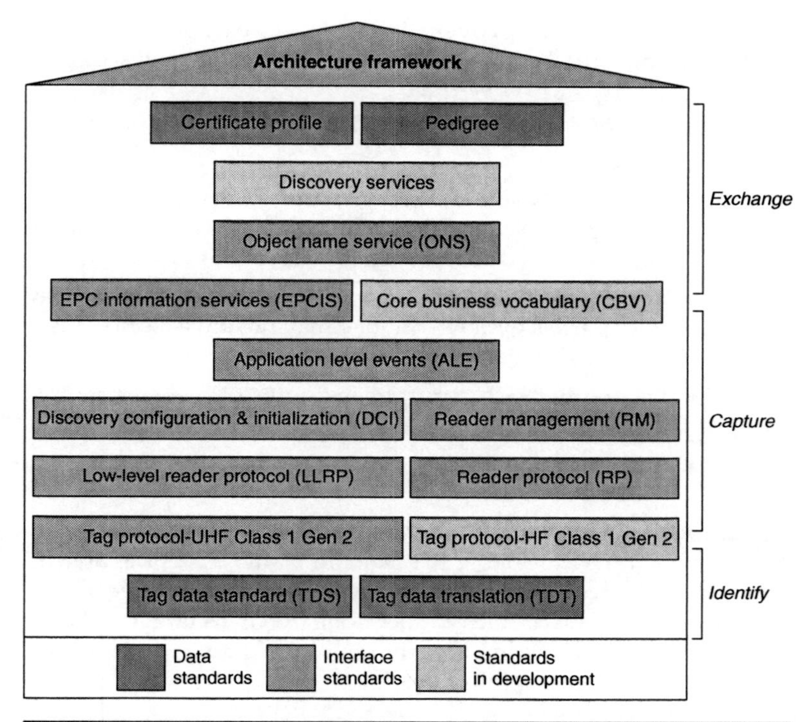


**FIGURE 3.2** EPCglobal architectural framework. (*Source*: http://www.epcglobalinc.org/standards)

defined in this document are based on two Internet standards, defined by the Public Key Infrastructure (PKIX) Working Group of the Internet Engineering Task Force (IETF), that have been well implemented, deployed, and tested in many existing environments.

### Pedigree Standard

This document and its associated attachments specify an architecture for the maintenance and exchange of electronic pedigree documents for use by pharmaceutical supply chain participants. The architecture is designed to aid compliance with document-based pedigree laws.

### Discovery Services Standard (In Development)

The Discovery Services Standards is currently in development by the EPCglobal Data Discovery Joint Requirements Group.

> *What is "discovery"?* "Discovery" is finding and obtaining all relevant visibility data that a party is authorized to access, when some of that data is under the control of other parties with whom no prior business relationship exists.

*Why is it needed?* Discovery is needed to exchange information along broader and different pathways than traditional e-commerce.

*What will be the benefits of a discovery services standard?*

- Enable trading partners to discover all of the resources who may have information about things (Who has data about this specific EPC? Where is their EPC Information Service located so I can ask for data about this EPC?)
- Enable trading partners to exchange data in a secure way with parties with whom they may not have a prior direct business relationship
- Ensure that each party retains rights of ownership of its visibility data
- Ensure that queries are authorized and authenticated

### Object Naming Service (ONS) Standard

This document specifies how the Domain Name System is used to locate authoritative metadata and services associated with the SGTIN portion of a given Electronic Product Code (EPC). Its target audience is developers that will be implementing Object Naming Service (ONS) resolution systems for applications.

### EPCIS: EPC Information Services Standard

This document is an EPCglobal normative specification that defines Version 1.0 of EPC Information Services (EPCIS). The goal of EPCIS is to enable differing applications to share EPC data, both within and across enterprises. Ultimately, this sharing is aimed at enabling participants in the EPCglobal Network to view the disposition of EPC-bearing objects within a relevant business context.

### Core Business Vocabulary (CBV)

The goal of Core Business Vocabulary (CBV) is to specify the structure of vocabularies and specific values for the vocabulary elements used in conjunction with the Electronic Product Code Information Services (EPCIS) standard for sharing event data both within and across enterprises. The aim is to standardize these elements across participants in the EPC Network to improve the understanding of data contained in EPCIS events.

## Capture

### Application Level Events (ALE) Standard

This EPCglobal board-ratified standard specifies an interface through which clients may obtain filtered and consolidated EPC data from a variety of sources.

### Discovery, Configuration, and Initialization Standard for Reader Operations

This GS1 EPCglobal standard specifies an interface between RFID readers and access controllers and the network on which they operate. The purpose of this document is to specify the necessary and the optional operations of a reader and client that allow them to utilize the network to which they are connected to communicate with other devices, exchange configuration information, and initialize the operation of each reader, so that the Reader Operations Protocols can be used to control the operation of the readers to provide tag and other information to the client. To facilitate these operations by the reader, an access controller provides several functions, described below.

The following are the responsibilities of this interface:

- Provide a means for the reader to discover one or more access controllers
- Provide a means for the access controller to discover one or more readers
- Provide a means for the reader to discover one or more clients
- Provide a means for the reader and access controller to exchange and authenticate identity information
- Provide a means for the client and access controller to authenticate their communications and operations
- Provide a means for the access controller to configure the reader, including a means to update the software and/or firmware on the reader
- Provide a means for the access controller to initialize the reader, providing parameters necessary for the reader to begin operation
- Provide a means for the reader and access controller to exchange vendor-specific information

The target audience for this specification includes:

- RFID network infrastructure vendors
- Reader vendors
- EPC middleware vendors
- System integrators

### Reader Management (RM) Standard

The current RM Standard Version 1.0.1 of the wire protocol is used by management software to monitor the operating status and health of EPCglobal-compliant RFID readers. This document complements the EPCglobal Reader Protocol Version 1.1 specification [RP1]. In addition, this document defines Version 1.0 of the EPCglobal Simple Network Management Protocol (SNMP) RFID management information base (MIB). Version 1.0.1 corrects errata discovered in Version 1.0.

#### Low-Level Reader Protocol (LLRP) Standard

This document specifies an interface between RFID readers and clients. The interface protocol is called low level because it provides control of RFID air protocol operation timing and access to air protocol command parameters. The design of this interface recognizes that in some RFID systems, there is a requirement for explicit knowledge of RFID air protocols and the ability to control readers that implement RFID air protocol communications. It also recognizes that coupling control to the physical layers of an RFID infrastructure may be useful in mitigating RFID interference.

#### Reader Protocol (RP) Standard

Reader protocol is an interface standard that specifies the interactions between a device capable of reading/writing tags and application software.

### Identify

#### Class-1 Generation-2 UHF Air Interface Protocol Standard: "Gen 2"

Commonly known as the "Gen 2" standard, this standard defines the physical and logical requirements for a passive-backscatter, interrogator-talks-first (ITF), radio frequency identification (RFID) system operating in the frequency range of 860–960 MHz. The system comprises interrogators (also known as readers) and tags (also known as labels).

The UHF Class-1 Generation-2 air interface protocol V1.2.0 extends the item-level tagging capabilities of UHF Gen 2. In this protocol, three optional features have been added: (1) An indicator is now available to show when there is formatted data in user memory. (2) Addition of permanent locking (permalocking) on a block level in user memory now protects contents that have already been written. (3) Recommissioning of a tag after point-of-sale (POS) operations is now available. The recommissioned action is indicated through the inclusion of extended protocol control bits.

#### HF Generation-2 Tag Protocol Standard (Coming Soon)

This specification was developed by the Hardware Action Group (HAG) HF Air Interface Working Group. The specification is currently a proposed specification in legal review.

#### EPC Tag Data Standard (TDS)

This standard defines standardized EPC tag data, including how it is encoded on the tag and how it is encoded for use in the information systems layers of the EPC systems network. This version is only applied to tag types in common use at the time of publication and does not provide specific guidance on Gen 2 tags. It includes specific encoding schemes for:

- **GTIN**: GS1 Global Trade Item Number(here, a serialized version)
- **SSCC**: GS1 Serial Shipping Container Code
- **GLN**: GS1 Global Location Number
- **GRAI**: GS1 Global Returnable Asset Identifier
- **GIAI**: GS1 Global Individual Asset Identifier
- **GID**: GS1 General Identifier
- **GDTI**: GS1 Global Document Type Identifier
- **GSRN**: GS1 Global Service Relation Number

### EPC Tag Data Translation (TDT) Standard

The EPC Tag Data Translation standard is concerned with a machine-readable version of the EPC Tag Data Standards specification. The machine-readable version can be readily used for validating EPC formats as well as translating between the different levels of representation in a consistent way. This specification describes how to interpret the machine-readable version. It contains details of the structure and elements of the machine-readable markup files and provides guidance on how it might be used in automatic translation or validation software, whether stand-alone or embedded in other systems.

Version 1.4 of the TDT specification is fully compatible with TDS Version 1.4.

## An Example: GS1 EPCglobal's RFID-Based EAS

RFID-based electronic article surveillance (EAS) is a technological method for deterring and detecting theft of consumer goods. RFID-based EAS tags (based on the EPCglobal Gen 2 standard) are fixed to an item's packaging or to the item itself. These tags can be removed and/or disposed of by consumers or sales associates after purchase. The goal of RFID-based EAS is to combine the known benefits of RFID, such as increased supply chain visibility and improved inventory tracking and process productivity along the supply chain including the retail sales floor, with the advantages of an EAS system (item-level theft deterrence, detection, and protection).

GS1 EPCglobal's RFID-based EAS Phase 1 Group developed a set of common retailer requirements for using RFID-based EAS. Most of these requirements are fulfilled using current standards for disposable and/or reusable tags. Postpurchase-disposable tags can be altered by the retailer. They are generally removed and discarded by the consumer or retailer and include the following:

- Fabric: pouch with RFID-based EAS device enclosed and sewn on garment
- Hang tags: RFID-based EAS device integrated into paper hang tag (swing ticket) or pocket flasher
- RFID-based EAS device integrated into a self-adhesive label
- Drop-in tags: RFID-based EAS device that is dropped into a pocket of a garment, which is sometimes then stitched up
- Plastic: RFID-based EAS device embedded or encased in plastic; for example, plastic hanger, integrated seal, or disposable hard tag

For example: Single-use tags are small, lightweight, hard tags intended for one-time use, which are removed at the point of sale and discarded.

Reusable tags, generally applied by the retailer, retailer's supply chain, or supply chain partners, are removed at the point of sale. They include the following:

- Hard tags: RFID-based device encased in plastic housing, with a secure method of application

Within phase 2 of this RFID-based EAS group, an implementation guide is written to take advantage of the current standards to provide general implementation guidelines for others who may wish to deploy an RFID-based EAS system. The solution depends on using either a reader with a simple database or access to a network database to determine whether an item has been sold or not. The EPCglobal RFID-based EAS Technical Implementation Guide provides guidelines on how to technically implement RFID-based EAS using current GS1 and EPCglobal standards. The retailer needs to realize that with this approach if a reader fails or the database goes down, then RFID-based EAS functionality would be compromised. This loss is not obvious to a customer. Information for products sold during this time can be captured so that the product can later be removed from the database.

With legacy EAS systems, the tags performed only the primary purpose of deterring and detecting theft. They were separated, isolated systems with a single function. Previously, retailers could have various tags applied including ones from the manufacturer, ones from the retailer, and the additional EAS tag. Also, due to competing Internet protocol (IP)-protected technologies, EAS tagging could lead to multiple inventories for manufacturers to suit the needs of individual retailers. As a result of this complexity, less source tagging occurred.

With a collective approach, RFID-based EAS can also be used to help with inventory, returns, detection of counterfeit products, and much more. A key advantage with an RFID-based EAS system is that item-level visibility could be available at the point of exit and entry (PoE). With legacy EAS systems, no actionable intelligence is provided when the alarm sounds; the only information that is known is that an EAS tag has passed the pedestals. With RFID-based EAS, when an alarm is activated, there is information not only that an item has passed the pedestals but also which particular item has activated the alarm and the quantity of items passing through the pedestals at that moment. This timely information will help the loss prevention department refine its strategy concerning the deployment of theft deterrent resources. This solution will allow a manufacturer to apply one tag that can be used throughout the supply chain for multiple functions. All costs of readers, tags, hardware installation, application of tags, and so on could be dedicated to the single-tag approach. Another advantage of a common implementation guide is the ability to leverage existing public policy work within EPCglobal. To encourage consumer acceptance, it is advisable for the retailer to adhere to the EPCglobal Consumer Guidelines available at http://www.epcglobalinc.org/public/ppsc_guide/.

## Source Tagging

RFID-based EAS tagging will be based on open standards, leading to a reduction of multiple inventories for those manufacturers who previously needed to use different tags for various retailers. It will lead to more items being secured by EAS functionality from the source. Utilizing the serialized information available via RFID will improve supply chain visibility and help prevent and detect shrinkage throughout the logistics chain and the retail environment. One of the goals of the implementation guide is to increase source tagging by having a common standard. Source tagging is defined as the application of RFID-based EAS security tags at the source, the supplier or manufacturer. For the retailer, source tagging eliminates the labor expense needed to apply the RFID-based EAS tags themselves, and potentially reduces the time between receipt of merchandise and when the merchandise is ready for sale. For the supplier, the benefits include the opportunity to use RFID for inventory management and visibility. In addition, it allows the option to preserve the packaging aesthetics by incorporating the tags within the product packaging. Source tagging allows the RFID-based EAS tags to be concealed and more difficult to remove, if desired.

In summary, the implementation guide shows the retailer how to use RFID-based EAS functionality based on current GS1 EPCglobal standards. The implementation guide concentrates on goods receiving, points of entry and exit, and points of sale. These points of entry and exit can be retail store doors, break rooms, restrooms, and so on. Upon receipt of goods, an inventory check occurs with all items' serialized EPC numbers added to a database. Upon the sale of an item, the number is removed from the database prior to its leaving the store. At the exits of the store, a detection system sounds an alarm or otherwise alerts the staff when it senses tags that have not been removed from the inventory database.

# PART II

# Issues and Challenges



Part Two of the book addresses issues in deciding whether to adopt RFID and challenges in designing a system. Chapter 4: *Challenges in Designing RFID Applications* looks at the primary issues for designing RFID applications, which involve operational, technical, and, of course, financial challenges. In addition, the security and privacy challenges are introduced as well as the growing social concerns involving environmental issues.

Chapter 5: *RFID Security and Privacy* provides an overview of RFID security and privacy concerns, followed by implications of these threats.

The challenge for companies interested in the adoption of RFID is to first make a business case that this technology would be in the long-term interest of the user. The decision to adopt RFID should be made on its ability to provide value to the organization. Chapter 6: *Business Analytics* examines the factors to consider for adopting RFID systems.

# CHAPTER 4

# Challenges in Designing RFID Applications

Increased popularity does come with challenges (some common challenges are shown in Fig. 4.1). While cost and standards have been known as major barriers for adoption, several implementation challenges also exist. These challenges in designing an RFID system are categorized as operational, technical, financial, security and privacy, and environmental (Fig. 4.2).

## Challenges

### Operational Challenges

To meet EPCglobal standards, companies are being required to label cases and pallets, and sometimes individual items. As more individual items are being tagged, the challenge becomes positioning tags so they can be read within a case or pallet. As more and more individual items are being tagged, there will also be a need to automate the tagging process because of the huge volume of individual items.

Some rate of failure to read can happen. This is especially true with metals and liquids. In the pharmacy supply chains, for example, tagging of liquids and biological materials will be challenged due to the space issues on the exterior of bottles and the questionable effect of radio waves on these products.

In hospitals and health care services, RFID systems have the potential for interference with other wireless communication devices.

### Technical Challenges

Technical challenges begin with the lack of consensus on standards. While standards already exist, establishing global standards is the real challenge. In the United States and Europe, EPCglobal is the standard.

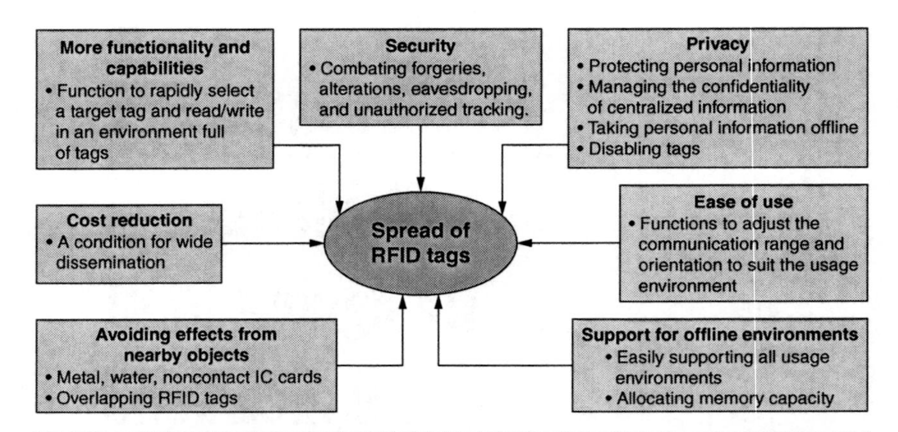


**FIGURE 4.1** Challenges of RFID adoption. (*Source*: http://www.frontech.fujitsu.com/imgv3/jp/group/frontech/en/forjp/rfid/rfidrw/interview/img01.jpg)

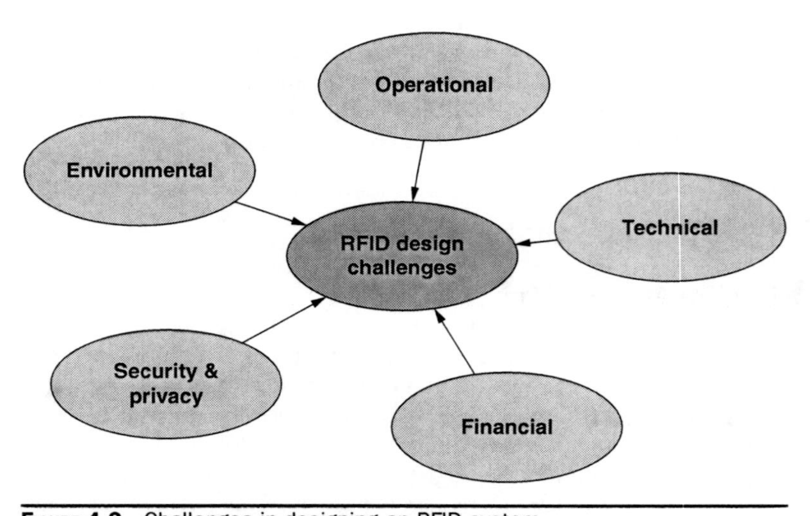


**FIGURE 4.2** Challenges in designing an RFID system.

Asian countries use their own classification system, such as the NPC (National Product Code) in China. Japan uses a different standard that does not communicate with EPCglobal standards.

Similar to the early years of enterprise resource planning (ERP) systems, the challenge of RFID implementation comes from the integration of RFID systems and the data they generate with other functional databases and applications. This challenge increases as companies integrate with supply chain partners.

### Financial Challenges

Cost is a major factor in determining the speed at which RFID technology is adopted. An RFID system requires expenditures not only for tags, readers, hardware, and software, but also for system maintenance and training. While consulting firms have estimated an investment of $13–25 million to implement RFID systems, it should be noted that costs are coming down. The costs of tags have steadily been decreasing from $1 per tag to around 10 cents (depending on volume purchases). The cost of readers and equipment has also decreased significantly.

Furthermore, cost concerns typically overshadow more important issues. For any RFID system to be successful, proper commitment, planning, and partnering are critical.

### Security and Privacy Challenges

RFID may pose a degree of security and privacy risks to both the company and individuals. For example, unprotected tags are vulnerable to potential eavesdropping, traffic analysis, and spoofing. Professional hackers could seek data for misuse, as well as the potential for corporate espionage—both of which are risks even without RFID systems.

RFID technology used at the item level could be used to gather information about customer behavior and potentially track their movements without their knowledge. Although some steps have been taken to reduce the consumer privacy concerns, if tags are being used at the item level, they should be deactivated (killed) at the point of sale. More on this challenge is provided in Chap. 5: *RFID Security and Privacy*.

### Environmental Challenges

As the tag prices come down and individual items being tagged proliferate, the tags also proliferate. RFID tags present potential environmental challenges because they are not biodegradable and may contain poisonous metals. A proposed solution is to set up reverse supply chains for the recycling or reuse of tags.

## RFID System Design

### Variables and Factors

The success of the RFID system depends largely on several variables. The variables and the factors involved in them need to be considered when designing and implementing an RFID system. These are explored in the following subsections.

### Operating Frequency

Operating frequency is the most important variable in designing the RFID system. This is the first question that the system designers must answer because the RFID system capabilities depend on the frequency type. The primary factors for determining the frequency type include read distance, application type, and operating conditions.

The read distance is simply the maximum distance between the reader and tag at which the tag data can be accurately read by the application. This maximum distance determines the frequency of the system. The read distance typically narrows down the range of frequencies available for use. For example, an RFID system (in the United States) requiring a close proximity read distance may use either 125 to 134 KHz (LF) or 13.56 MHz (HF). The read distance determines the frequency type. These frequency types are discussed in Chap. 2: *RFID* 101.

Characteristically, each frequency range is associated with the application type. For example, a close proximity frequency of 125 to 134 KHz (LF) matches the close proximity read application. More discussion on applications is provided later in this chapter.

Operating conditions can contribute to the operating frequency. For example, if the application needs to be less influenced by metal, then 125 to 134 KHz (LF) is a good choice. Similarly, in a hospital or health care environment where minimum interference with existing equipment is a requirement, 13.56 MHz (HF) is the appropriate choice.

### Objects to Be Tagged

This variable is among the most important and plays a significant role in a successful RFID system design. The factors that make up this variable includes the item types that will be tagged, the material and packaging type of the items being tagged, and how the item will be handled within the RFID system.

The item types that will be tagged can be categorized by level: individual item level, case-pack level, or pallet level. The individual item level is perhaps the most complex and costly. While a common conception of individual item level is consumer-packaged items in supply chain management, individual item level may also apply to tracking of assets, capital goods, or work in progress. The case-pack level is a collection of individual items, and the pallet level consists of a collection of cases.

The material and packaging type of the items being tagged should be considered. For example, tagging items made of RF-lucent material (such as plastics and paper) is much easier when compared to RF-absorbent (such as items containing fluids) or RF-opaque (such as metals). RF-absorbent material could lead to low tag reliability. If the material and/or packaging contain metal (RF-opaque), it becomes difficult to get accurate reads.

Depending on the output power and duty cycle (discussed later in this chapter) of the readers, the RF energy may not reach all of the tags if an RFID tag is located at or below the depth reachable by that RF energy. A prime example is the number of cases placed on a pallet. Even if the correct RFID tag is used for the material and packaging type, it is still possible for some case-level tags to fail to be read because the RF energy was not able to penetrate to that depth.

The opposite can also lead to failure. In this case, the pallet cases are small and the tags are so close together that they could couple the antennas of other tags and not be read. This is known as the shadowing effect.

How the item will be handled within the RFID system can also affect the readability of the tag. For example, a forklift can reflect and prevent the RF waves from reading the tags. Use of wireless devices in forklift handling could also interfere with the RFID system. And of course, the speed at which the forklift moves could also reduce the reliability of tag reading.

### Operating Environment

The operating environment is another variable that can greatly influence RFID system success by affecting how the RFID application will work. In general, the operating environment consists of temperature (extreme cold or heat), moisture, static, and shock/vibration levels. The operating environment may also contain obstructions and interference. RF waves could be blocked due to RF-opaque objects or mobile equipment. The operating environment may also include people, made up mostly of liquid, which is RF-absorbent and could dampen and weaken the RF wave.

### Tags

The factors that need to be considered for this variable are tag type, data capacity, readability, tag attachment, physical properties, and the volume of tags needed. This variable decision will affect the recurring application cost as well as the data collection capabilities. Finding the optimum balance among these factors may be difficult, and therefore acceptable trade-offs may need to be sought.

The tag type refers to the type of tag that will be used in the application (see Fig. 4.3). In general, the tag type is characterized as (1) passive, active, or semiactive, (2) read-only (RO), write-once, read-many (WORM), or read/write (RW), and (3) having special characteristics, if any. Typically, if the application does not require custom features, passive tags will work in the RFID system. However, if onboard processing such as sensing temperature and humidity is required, then active tags are needed. Semiactive tags are needed if the application involves high-speed movement of tagged objects. If the tag needs to store a unique static identifier, then a RO or a WORM is needed. However, if the tag contains dynamic data, then RW tags will be

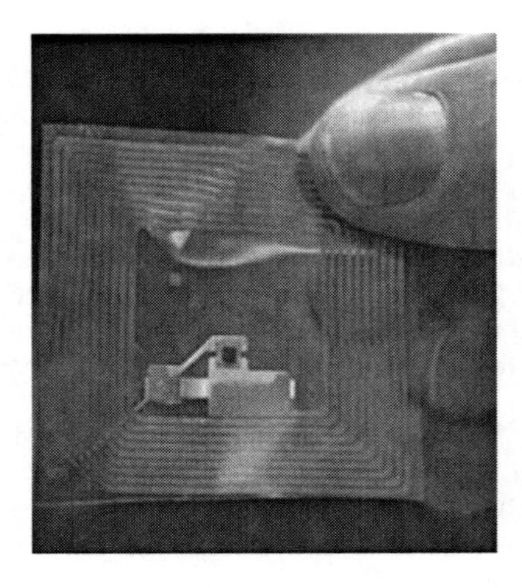

**Figure 4.3** RFID tag in use by Wal-Mart.

needed. Finally, some application types may require special characteristics such as customized tags to handle metallic objects or items that carry RF-absorbent liquids.

Data capacity is a factor that affects application cost and efficiency. Capacity largely depends on how the tag data is being used: as a "license plate" or as a tag that contains object attributes. For a "license plate" type, a 96-bit EPC Class 0+/1 or even a Class 0 is common for the unique identifier. If information on a specific attribute of the tagged object is needed, such as part number and maintenance history, a larger memory size (such as 256 bits or more) is needed.

Readability is an important factor associated with the tag—this factor alone could determine the success of the RFID system. Tag robustness refers to how many times a tag can be read by the reader. Therefore, the more robustness, the better the probability that the tag would be read when interrogated. The antenna design influences the tag readability. Tag density indicates how many tags are present in the read field; the greater the tag density, the lower the robustness and the lower the reliability of the tag. Tag orientation also plays an important role in readability and depends on the antenna type being used. The tags have to be oriented to the antenna so that it can be aligned with the electromagnetic field of the antenna. Of course, the speed at which the tagged object moves can also affect the read reliability.

Where and how the tags are attached also plays an important role in read reliability. Experimentation is necessary because there is not a single answer for all objects. Good practice for tag placement is near the UPC label of the tagged object so it can be easily located. How the tags are attached will depend on the object. A simple adhesive back strip is sufficient to mount tags on products made of RF-lucent material. For products made of RF-absorbent or RF-opaque material, the mounting of the tag becomes more complex. Typically, a spacer made of RF-friendly material can be used between the object and the tag.

Physical properties include the tag dimensions and durability. Tag dimensions may depend on the size and shape of the tagged objects, or more specifically on the "real estate" available for a tag on the object, meaning that available space may require a much smaller tag than expected. The tag's durability essentially depends on the environmental operating conditions of the RFID system. Some of these environmental conditions may include heat, cold, humidity, or mechanical vibrations.

The volume of tags needed for the RFID application is another concern. The cost will be part of the recurring overhead to operate the RFID application, and may be reduced if the tags can be recycled.

### Readers

The reader is the central point for collecting data. This is why most of the RFID system's data collection capabilities are directly influenced by the readers. The main factors to consider are reader type, upgradability, manageability, and durability.

The reader type must first support the operating frequency. Ideally, a multiple-frequency reader is desired. Next, the reader should support multiple antenna ports. For example, a four-port reader has been known to work better in terms of read zone coverage compared to a two-port reader. Another factor to consider is the support for the input/output (I/O) ports needed by the application. The readers might also need to support interfaces for sending tag data, I/O control, monitoring, and management, and depending on the application, specialization may be required (like a motion sensor). Finally, the application may require handheld readers or stationary readers.

The reader is a hardware investment, and therefore it is wise to plan for upgradability for future development. It is strongly recommended that any reader, whether handheld or stationary, should be upgradable. Ideally, the readers should be manageable remotely so that their use can be centralized. For the most part, the readers are typically durable. But a stationary reader could be positioned in an enclosed housing to prevent damage or influence by the operating environment.

### Antennas

Basically there are two types of antennas. Depending on the RFID application, circular polarized or linear polarized antennas are used. For example, if the orientation of the tags can be arbitrary to the antenna, then a circular polarized antenna should be used. However, if the orientation of the tags cannot be arbitrary (i.e., a fixed orientation relative to the antenna) or if a long read range is needed, then a linear polarized antenna should be used. For the most part, the antennas are typically durable. But the antennas could also be enclosed in a housing to prevent damage or influence by the operating environment. The RFID system design should consider the antenna footprint, power and duty cycle, and antenna installation.

The three-dimensional area through which the RF wave radiates from the antenna attached to a reader is referred to as antenna footprint. The RFID tags in that footprint can be read by the reader. The actual shape of the footprint depends on the operating environment.

The power level and the duty cycle (the fraction of time the reader emits energy) influences the antenna footprint, and cannot be increased without permission from government regulatory authorities. However, they can be reduced. Such a decrease could prevent the RFID system from interfering with neighboring RFID systems (typically found in access control applications).

Antenna installation can be a challenge. The antenna should be placed away from metals, but close enough to where the tags will enter the antenna footprint. For maximum antenna footprint, test the height and antenna orientation to determine the appropriate placement. Employee safety should also be taken into account.

## Vendors and Suppliers

An RFID system is a technology investment and regardless of the application, one variable will always be important: vendor selection. Much can be said about the criteria for vendor selection. The most important factor is to carefully evaluate multiple vendors. It is this author's recommendation that vendors should be invited to participate in the RFID system design, and each evaluated based on testing and test cases.

It is highly recommended that factors in vendor selection include vendor support, plans for upgrading, and contingency plans. RFID systems today are much different from those implemented just a few short years ago. Vendors are always updating their hardware and software; hence vendor support and how upgrades are handled can be critical to continued RFID system success. While establishing vendor relationships is important in any technology adoption, contingency plans should be in place just in case the vendor can no longer support changes to the RFID technology (or worse: the vendor decides to leave the RFID industry).

## Systems Integration

The integration of the RFID system with other existing systems may or may not be necessary (at first). However, the RFID system is primarily a data-collecting system. At some point, the RFID system will need to be integrated with existing enterprise resource planning (ERP) or warehouse management systems (WMS), both of which have integration capabilities.

# CHAPTER 5

# RFID Security and Privacy

Like many new technologies, RFID systems may create security and privacy challenges. First, unprotected tags may make an organization vulnerable. The organization's security may be at risk and opened to eavesdropping, spoofing, traffic analysis, and misuse. Corporate espionage is the primary risk, because unprotected tags could be monitored and tracked by competitors. Hackers could also tamper with the tags and reduce the price of expensive items and use self-checkout to avoid store clerks. These challenges could be corrected with programmable read / write tags encoded with security features that limit access, as well as identifying and recording unauthorized access.

Consumer privacy concerns have also surfaced. In this case, RFID used at the item level could be used to obtain information about customers, and to track their movement without their knowledge (Fig. 5.1). However, tags used at the item level could be deactivated at the point of sale. Tags could also incorporate blocking and encryption systems designed to protect privacy and unauthorized access. The general public will require education in order to overcome this challenge.

Major companies worldwide have scrapped RFID programs following consumer backlash (Fig. 5.2), and several U.S. states including California and Massachusetts are considering whether to implement RFID-specific privacy policies. While radio frequency systems are highly regulated by government agencies to ensure the technology poses no serious health concerns to the general public, the issue of consumer privacy for millions of individual consumers worldwide is indeed real. However, several projects have been embraced by the public without any resistance. For example, a number of U.S. highway toll payment systems, including the E-ZPass developed by a consortium of northeastern states and Exxon Mobil Corp.'s Speedpass system for credit card transactions, have been accepted and grown into successful businesses.

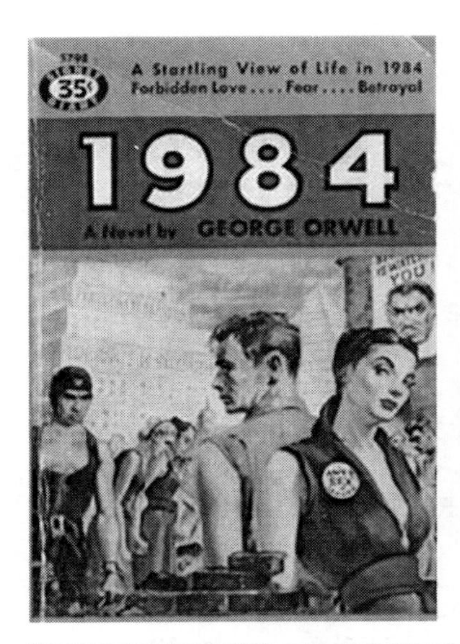


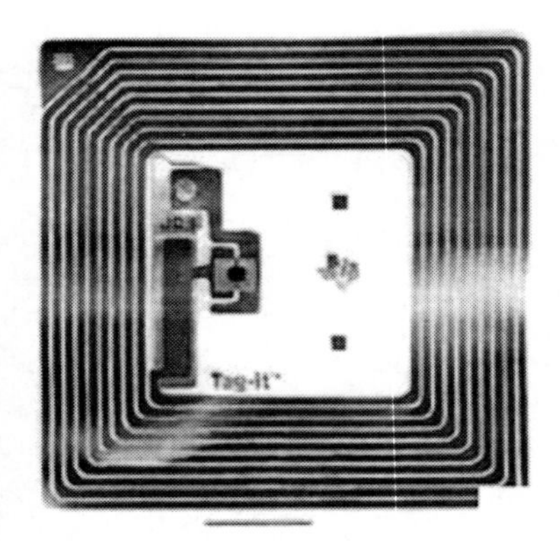


**FIGURE 5.1** Big brother and the thought police. (*Source*: http://dvice.com/pics/1984rfid2_w.jpg)

**FIGURE 5.2** RFID "spychip" protest. (*Source*: http://www.spychips.com/images/nyc-protest-8-13-08.jpg)

The key to a successful RFID application is to consider the equation from the consumer's point of view. Consumers accept the risk of being tracked and their activities being monitored if they feel it is worth the benefits the application provides. Companies introducing RFID to track inventory levels and manage replenishment in thousands of retail stores need to view their RFID system benefits and risks from the consumer's point of view. Unless companies are able to tie the presence of tags directly to consumer benefits in terms of reduced price, better products, or quicker checkout, they will likely continue to see more resistance to the technology.

Any effort to implement RFID in the retail environment can be improved in two basic ways. First, lower the risk to consumers of losing their personal data and privacy. Second, increase the benefit to consumers in terms of a more convenient shopping experience, lower prices, and quicker checkout. RFID developers have sought to limit the perceived risk by trying to educate consumers about the benefits of RFID and providing privacy policies to explain what data is being collected and how it's being used.

In order to increase consumer acceptance of RFID technology, RFID advocates must promote and implement comprehensive security measures along with consumer education, enforcement guidelines, and research in and development of practical security technologies.

## Security Implications and Privacy Threats

### RFID Security

Consider the issuance of passports and driver's licenses with RFID chips. Privacy is not the only issue here: Researchers say that unauthorized reading would threaten border security as well. If it is easy to get the identification number from the cards, then it's relatively easy to counterfeit them, simply by loading a stolen ID number onto a blank off-the-shelf chip. If each RFID chip also had a unique, hard-wired serial number, which had to correspond to the stored ID number, it would be hard to counterfeit.

For example, Washington State's Department of Licensing began using EPC Gen 2 UHF tags embedded in the state driver's license to improve traffic flow across the Canadian border with the approval of the U.S. Department of Homeland Security. The RFID-enabled driver's license is an alternative to showing a U.S. passport at the border crossing.

But neither the Washington State driver's license card with an optional RFID tag nor the passport cards have that extra security feature.

The Washington cards are open to one additional type of attack: EPC tags can be disabled when a reader issues a "kill" command. Although each tag is designed to be protected by a PIN that allows only authorized users to issue the command, the state never set the PIN on the cards it distributed, allowing anyone with the RFID reader to set it himself and commence killing cards. If a good number of Washingtonians with enhanced licenses were gathered at a border crossing, someone could cause a disruption by killing large numbers of cards. An attacker could also use this tactic to harass particular individuals, since a killed card is likely to draw suspicion.

RFID is a wireless technology and is therefore subject to third-party interception unless the signal is secured. Today's RFID tags can incorporate only a restricted amount of security. This is largely due to the limited amount of data that they can hold, and presently tags do not employ passwords or read access control into a secure or designated area.

But the real threat comes from vulnerability of the firm's databases. Hence, the key aspect of RFID security is to protect the firm's database and customer information. Protecting this information so that people cannot hack into the system is the same challenge as for Internet systems. System reliability is an equally important concern due to the possibility of the data being compromised or altered by an unauthorized resource.

Threats to corporate data security can be summarized as follows:

1. *Corporate espionage.* Tagged objects in the supply chain make it easier for competitors to remotely gather supply chain data, which are industry's most confidential data.
2. *Competitive marketing.* RFID systems may make it easier for competitors to gain unauthorized access to customer preferences and use the data in competitive marketing scenarios.
3. *Infrastructure.* RFID systems make organizations susceptible to new kinds of denial of service attacks.
4. *Trust perimeter.* As organizations increasingly share large volumes of data electronically, sharing mechanisms offer new opportunities of attack.

## RFID Privacy

The major risk of the widespread implementation of RFID is the possibility of privacy violation. RFID tags are known to transmit information to the readers without the knowledge of the bearer, which has severe implications in case of sensitive data. For example, if personal identity were linked with unique RFID tag numbers, individuals could be profiled and tracked without their knowledge or consent. Several areas of concern include the following.

*Personal data protection.* For example, if information is gathered about purchasing habits, this information could be later used by insurance companies to charge higher premiums if healthy diets are not purchased. In general, purchasing information could be used for data mining for profiling purposes. Yet a more pressing concern than the publicly available information is the increasing collection, analysis, use, and sale of nonpublic, personally identifying information from commercial entities.

*Tracking and tracing using EPC-compliant RFID tag.* Another concern is the consumer's physical privacy. If the tag stays active after the consumer has left the store, a company may be able to track the physical movements of the consumer as far as prevailing technology permits. If a company chooses to sell its database of consumer information, other companies could track that consumer's movements as well.

*Secret tag reading.* RFID tags and chips can be well hidden inside product packaging or items of clothing, apparel, books, and electronic devices. RFID readers can also be concealed. In that way, information contained by the RFID tags can be readily obtained without the individual's knowledge.

The use of RFID technology can have profound consumer privacy and civil liberty implications. For example, a driver whose car has an E-ZPass RFID tag slows down to pass through a gate that reads the tag so that the highway toll can be deducted from a prepaid account, but does not have to stop to pay. While this does provide a level of convenience, this same technology could also show how fast the driver was going between the collection points. This opens up questions of privacy.

In another example, books are checked out at a library. The privacy question may be who is able to monitor that information once the patron leaves the library. The library RFID reader could also be used to scan you for all of the purchases you are carrying, which could be used to further understand your buying habits. This is not dissimilar to how grocery stores currently collect information on purchasing habits with their reward cards.

## Solutions to Security and Privacy Risks

Possible solutions to security and privacy concerns include defensive measures that could be taken in three areas: RFID tags and readers; policies for securing an organization's information; and information system technology.

Defensive measures that could be incorporated into tags and readers include proposed solutions as well as some that are already developed:

1. *Kill command.* Kill the tags upon purchase of tagged product. However, as a widespread solution, killing tag functionality at sale curtails the future potential use of RFID in consumer services (such as in smart refrigerators that automatically reorder food products, tags that give an alarm at expiration date or product recall, and systems for personal library management). It would also prevent the use of RFID information when products are resold or recycled.

2. *Faraday cage approach.* Protect the tag by using a faraday cage, a container made of metal mesh that is impenetrable by radio signals of certain frequencies.
3. *Active jamming approach.* Use a radio frequency device that actively sends radio signals to block the operation of any nearby RFID readers.
4. *Blocker tag approach.* The tag replies with simulated signals when queried by a reader, so that the reader cannot trust the received signals.
5. *Rewritable memory.* The tag stores an anonymous ID, so that an adversary may not know the real ID of the tag.
6. *Symmetric key encryption.* The tag and its legitimate reader use an authentication mechanism based on a simple two-way challenge-response algorithm.
7. *Public key encryption.* The tag information is encrypted based on the cryptographic principle of reencryption.

Defensive measures that rely on good security methods:

1. Establishing detailed policies for the security of the data
2. Assessing the value of the asset being protected
3. Incorporating security solutions that are transparent
4. Viewing security as a process
5. Realizing that security is an ongoing process

Defensive measures include privacy-enhancing technology for sharing information:

1. Virtual private networks
2. Transport layer security
3. DNS security extensions
4. Onion routing
5. Private information retrieval

# CHAPTER 6

# Business Analytics

Building a business case for RFID is a challenging task, given that most companies already have auto ID technology (like bar codes) that are working. However, with each technology advance, whether technology push or technology pull, companies are always on the lookout for the next technology to give them a competitive advantage (Fig. 6.1). This chapter looks at different methods for evaluating the RFID system.

## Investing in RFID

Building a business case for investing in RFID can seem like an overwhelming challenge. Identifying opportunities to use RFID to improve information richness and visibility requires careful analysis. If implemented appropriately, RFID can benefit a business in time and cost savings. Most companies that have been successful in implementing RFID have adjusted their processes to capitalize on benefits of the technology. For example, Kimberly-Clark's philosophy for implementing RFID was based on redesigning processes to take advantage of the technology. Conversely, most companies apply the slap-and-ship method without changing their business processes. But the bottom line for all companies is that no technology can correct a broken or bad process. The process has to be fixed first.

To determine the most appropriate manner to implement RFID or even analyze whether RFID can be used to improve a process (or system), the current process needs to be mapped out. This can be done simply by constructing a flowchart map of the process. Once the current process is mapped out and analyzed, the process may need to be redone (reengineered) to implement RFID for its maximum benefit.

The analysis includes a feasibility study that looks at the economic, technological, and operational feasibility. Economic feasibility involves analyzing the cost of implementing the technology, the benefits, and even the cost of not implementing the technology, which is a major concern and must be at the forefront of any feasibility study. The technological aspects of the process along with RFID technology capabilities and shortcomings need to be considered for each application. Finally, the operational feasibility is the effect on both external

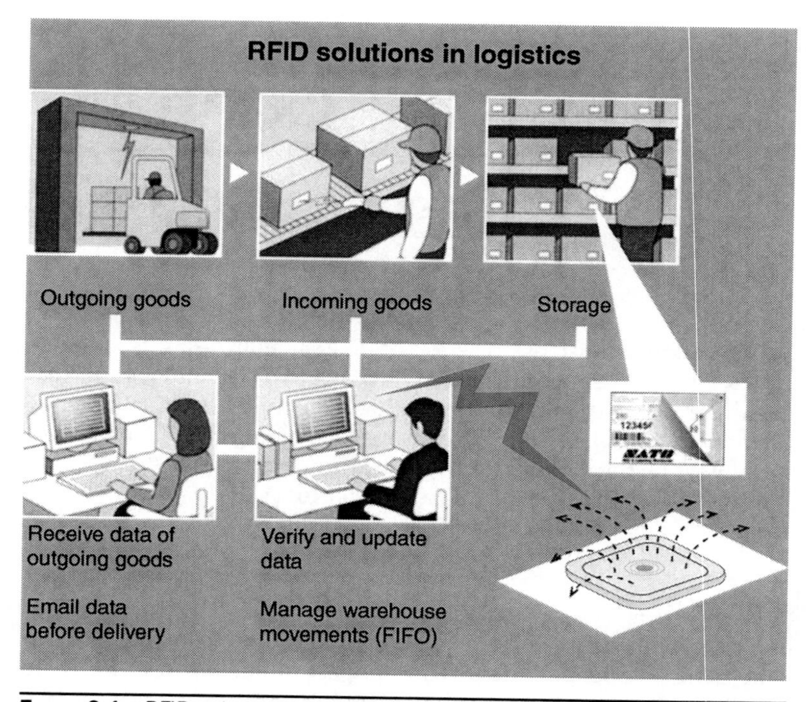


**FIGURE 6.1** RFID value in logistics applications. (*Source*: http://www.satoworldwide.com/images/RFID/RFID%20in%20Logistics%20Cartoon%20copy.jpg)

and internal customers. A system change requires managerial and employee support, including appropriate timing and scheduling for implementation of the technology.

## Examples of RFID Implementation Successes

After announcing the mandate that its top 100 suppliers use RFID on the pallet level in 2003, Wal-Mart reported that inventory accuracy improved by 13 percent, implying cost reductions and improved profits of as much as 10 percent. Wal-Mart is now mandating that most of its suppliers use RFID on the pallet level. RFID technology has the potential to change to the item level; 55 percent of all items could have RFID tags in the next 10 years.

Walgreens has implemented RFID tracking at its warehouse in Anderson, South Carolina. RFID tags are used to ensure that products are shipped to the correct retail store. The Anderson distribution center ships about 80,000 tagged totes daily to approximately 700 stores, with the RFID tags used to verify that totes containing products are loaded on the correct truck and in the correct order to ensure accurate delivery of goods.

The Department of Defense (DoD) reaffirmed its commitment to RFID technology in its Defense Federal Acquisition Regulation Supplement (DFARS 252.211-7006) in February 2007, mandating RFID tagging of any products sold to it. In this mandate, passive RFID tags have to be affixed to products at the packaging levels of case and palletized unit load, for shipments of items. The RFID initiative was started by the DoD in 2005 with the release of a Defense Federal Acquisition Regulation Supplement (DFARS) on September 13, 2005, by the Office of Management and Budget. In January 2009, the Government Office of Accounting (GOA) expressed concerns about the return on investment (ROI) for RFID in the DoD supply chain (Fig. 6.2). This, along with criticism by the GOA of weaknesses in the supply chain as far back as 1990, has resulted in a concerted DoD effort to revamp its supply chain. Process redesign will be an important part of improving the supply chain.

Another important application of RFID is in the pharmaceutical industry. Counterfeiting of drugs is a huge problem. To combat counterfeiting, RFID tags are used to track a drug from the factory to the retail pharmacy. This provides supply chain visibility to the pharmaceutical companies. All sorts of medical products, from syringes to

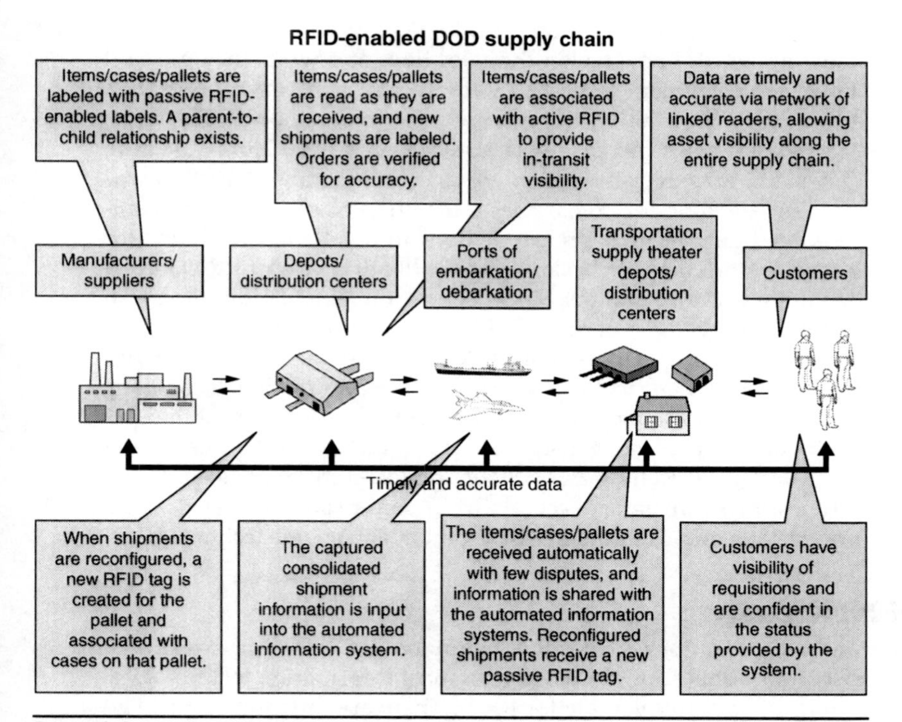


**FIGURE 6.2** RFID in Department of Defense supply chain.

ointments, are also tracked by RFID systems. Medical and pharmaceutical companies are trying to maximize the value of RFID to help minimize the risk of counterfeiting. When a company decides to go down the road to RFID implementation, a new set of vendors and integrated computer and software technology must be considered. Pharmaceutical companies, medical supply companies, hospitals, and labs are using RFID to better serve their customers and reduce their liability risk.

In the Las Vegas McCarran International Airport, RFID has improved baggage tracking by reducing lost baggage by 40 percent. Using RFID tags to track baggage would save airlines over $700 million per year; the costs would be recovered in one year, according to the International Air Transport Association. The cost to recover lost baggage is about $100 million annually. Improving baggage tracking would also improve customer service. At the Hong Kong International Airport, a reduction in handling cost from $7 to $4 per bag has resulted from the implementation of RFID tracking. Another factor in these cost savings is that RFID tracking systems have a reliability of more than 97 percent in read-rates compared to bar codes, which average only 80 percent.

American Apparel reinvented its management of inventory by attaching RFID tags to every item. This allowed managers to ensure that the correct inventory is on the sales floor available for customers. Knowing which items are on the sales floor and when they are sold allows replenishment to occur in an efficient and effective manner. When the RFID-enabled stores were compared to non-RFID stores, a 14 percent sales increase was noted. Labor reduction was 20 to 30 percent, inventory reduction was 15 percent, time reduction was noted in searching for items, sales transactions were faster, and complete inventory verification can be done in less than three hours for the entire store. The payback period for the first year of RFID implementation was four months. Since its pilot implementation, American Apparel is now implementing RFID in all of its 46 stores.

Another example of using RFID to reduce cost and improve return on investment (ROI) includes research at the Information Technology Research Institute at the Sam M. Walton College of Business at the University of Arkansas. This research was conducted in a 13-week study using item-level RFID tags at Bloomingdale's stores in 2008. Using RFID improved inventory accuracy by 27 percent and improved inventory tracking and counting by 98 percent over bar code technology.

## ROI for RFID

Many organizations are hesitant to implement RFID due to its initial cost and doubts about ROI. As with most information technologies, ROI is difficult to calculate for RFID. There are different ways to go about calculating ROI for information technology. Three approaches to determining ROI are described in the following sections of this chapter. These include business processes, business options, and cost-benefits analysis; combinations of these approaches are also discussed.

## Business Processes Approach

An in-depth look at business processes is done by analyzing those processes that will offer the most gain from RFID change. These processes are determined through interviews with supervisors and managers responsible for core business processes. Each task is analyzed for manual and automated processing to determine how the present process works. The next step is to determine how RFID could be implemented in the process and any changes necessary to allow RFID to be implemented. Processes often need reengineering in order to implement RFID for its full potential benefit.

Before processes are reengineered or changed, measures of throughput (amount of data processed) and cycle time (the length of time for a process of interest) should be recorded, and then projected throughput and cycle time should be estimated for RFID implementation. Often, simply identifying value-added and non-value-added tasks and eliminating the non-valued-added tasks that are unnecessary can improve throughput and cycle to a great extent. This may be a good time to implement these changes and get improvement without technology change.

After process throughput and cycle time are determined for the present process, then reengineering may be needed for RFID implementation. To determine reengineering needs for RFID, a complete analysis of each task and process must be done. This includes looking for data collection points needed to complete products or services and determining the need for data collected. Also, consultation with employees and supervisors to determine how data is collected and how it is used will be necessary to identify possible improvements in data collection and information flow that can be obtained through RFID implementation. Data collected in a process must be traced to the next level of use, where it can be determined if other data is needed from the process that may be obtained by implementing RFID.

This is a time consuming but necessary part of determining the ROI for RFID. From this analysis, the cost savings and improvement in information gathering can be determined. Some processes may not show any improvement to ROI. These processes need to be analyzed in relation to the processes that do show an improved ROI due to RFID implementation. It may or may not be advantageous to change the processes that do not show any improvement due to RFID, but it may be necessary to implement RFID even in those processes due to the need for consistency or for information in the overall system.

Three issues must be considered before RFID is implemented: (1) system costs including tags, readers, controllers, software, integration, and maintenance, (2) infrastructure within, and related to supply chain partners, and (3) suitability of RFID technology for the product or service of application. System costs will increase due to the number of tags and readers needed for the application. Also, the quantity of tags and readers will depend upon the information detail needed or desired. Redesign of the processes will need to be figured in as a cost.

Infrastructure related to integrating RFID technology with the organization's existing information systems will be another major issue in implementing RFID. Hardware and software investments will be necessary to make this happen. If RFID is only implemented within the organization without extension to supply chain partners, the gain in ROI could be minimal.

Suitability of RFID technology is another consideration. In cases where liquids or metals interfere with the RFID technology, then RFID may be of little or no value. Reading distance, number of tags per container to read, frequency of signal, and electronic interference are technology issues that have to be considered.

After considering the issues of cost, infrastructure, and technology, the organization or business unit can run a pilot study to determine ROI by comparing throughput and cycle time and richness of information collected to determine the benefit of the RFID application. If the ROI is small, then the decision makers will have to decide whether the implementation will benefit the organization. In some cases the richness of information obtained may justify the implementation of RFID. Of course, if the ROI is substantial with the RFID pilot study, then a full-scale implementation can be done.

## Business Options Approach

The options approach to determining the ROI for implementing RFID looks at the ROI not only at the present time, but also in the future. In general, business offers a number of options for RFID implementation: growth, flexibility, innovation and learning, waiting, and abandonment. Growth indicates the technology's capacity to grow into new uses in the future. Flexibility refers to using RFID tools for multiple uses. Innovation and learning create an atmosphere of gaining knowledge about RFID to improve and enrich data collection. Waiting is another option that can be used to decide whether the technology will be adopted universally by an industry or in the supply chain. Finally, abandonment allows an organization to walk away from the technology in the early stages of adoption before investment in the technology gets too extensive.

RFID can allow for more efficient collection of data. This will initially show some level of increased ROI due to efficiency. Next, the

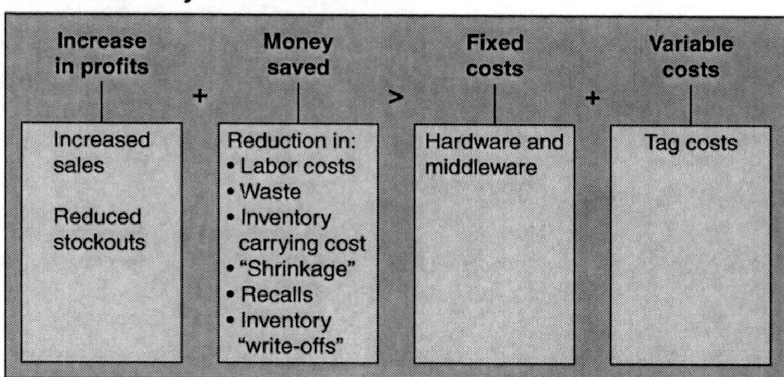


**FIGURE 6.3** Analyzing return on investment for RFID.

options of flexibility and innovation can be analyzed in light of the analysis of future benefits. These benefits may not be sufficient to allow for adoption. Waiting or abandonment may be the option to follow in this case. If flexibility and/or innovation show positive ROI, then adoption and extension of RFID should be followed.

## Cost-Benefits Analysis Approach

In the cost-benefits analysis approach, the costs of equipment and tags need to be calculated for implementation of RFID (Fig. 6.3). To determine this cost, the number of RFID readers and controllers must be determined along with the number and type of tags needed to collect the data required for the process under analysis. These costs can then be compared to projected benefits of using RFID. These benefits can be measured in time savings (labor reduction), improved information and data richness, and internal/external customer service. This approach does not take into account any process improvement issues or options.

## A Combination/Hybrid Approach

A combination of the above approaches could be used. In the cost-benefits analysis approach, a manager could look at various options after implementing a pilot process analysis using RFID. Or processes could be examined first and then RFID can be implemented using selected options.

A reasonable approach would be to analyze processes for value-added and non-value-added processes. Then processes are redesigned in light of the data needed from the system. This will take cooperation from managers, users, and customers to determine what data is needed from the system. The next step is to run a reengineered sample process with RFID in place and determine benefits in time savings, data richness, and internal/external customer service. On the basis of this initial analysis, options can then be executed, ranging from full-scale implementation to partial implementation to waiting or abandonment.

## ROI for Information Technology

ROI for any information technology like RFID must be evaluated not only with respect to hardware and software costs, but also with respect to installation, training, upgrades, service, ongoing management of the system, and system security costs. It may be advantageous to hire an RFID consultant to help identify technical requirements and resources needed to implement the system. Investing in RFID is a critical decision for companies. Whether and how it is implemented could mean a loss in competitive advantage for the company, resulting in loss of revenues in the long run. Careful analysis of ROI is vital for a company to make the most of its investment in RFID.

## Case Examples

### RFID Frees Up Patient Beds*

When nurses at St. Vincent's Hospital in Birmingham, Alabama, want to know if a patient has returned from having lab tests, they no longer need to walk to that person's room to check. They can simply look at a large, flat-panel screen hanging over the nurse's station, where an icon indicates the patient's location. The screen's color-coded graphics also tell them when lab results are ready and whether immediate attention is required.

Such patient-tracking and real-time clinical information is possible because of a system combining RFID technology and data from the hospital's various health-information programs. Nearly two years old, the system has helped the hospital improve capacity management and enhance the quality of its care. As a result, the hospital is able to manage the volume of patients better, with fewer bottlenecks in the process.

St. Vincent's is part of Ascension Health Corp., the country's largest nonprofit health care system, with 67 acute care hospitals in 20 states. Each year, St. Vincent's serves more than 17,000 inpatients and 125,000 outpatients. And the number of patients is growing: From March to December 2005, admissions jumped by 19 percent.

---

*Source: Jill Gambon. (2006). "RFID Frees Up Patients Beds." *RFID Journal*. Retrieved from http://www.rfidjournal.com/article/purchase/2549

But St. Vincent's, which has 338 beds, lacked up-to-the-minute information about the availability of the beds. As a result, patients had to be diverted to other hospitals. In 2004, St. Vincent's lost an estimated $20 million in net revenue because of such patient diversions.

To address this problem, the hospital developed a strategy to improve patient visibility, eliminate backups in admissions and discharges, and reduce the time spent on waiting for care. The first step in reaching those goals was getting better insight into where patients were at all times, as well as making real-time information available about the status of doctors' orders and test results. While the hospital was looking at how to make those improvements, it was also exploring ways to incorporate RFID into its infrastructure.

St. Vincent's has earned a reputation for embracing emerging technologies, including an expansive wireless local area network (LAN) covering more than 1 million square feet, and a computerized physician order entry system that aims to improve patient safety and boost the quality of care. After reading about the use of RFID by such early adopters as Wal-Mart and the U.S. Department of Defense, top management was convinced the technology held promise for the hospital.

Development work began in January 2004, and a pilot project was launched the following September in the 34-bed cardiac care unit. The tags were attached to the patients' charts that accompany them wherever they go in the hospital. The egg-shaped tags are slightly bigger than a key fob used to unlock a car. The system operates at 433.9 MHz and reads the tags every 10 seconds.

The RFID interrogators, wired into the hospital's Ethernet network, send information about the patient's location to an SQL Server database. Any location changes the interrogators detect are written to the database, then displayed in real time on screens installed throughout the hospital. To protect privacy, no names are displayed on the screens; only room numbers identify the patients.

In addition to the patient-location information, the system integrates clinical data and relevant information, such as notification of lab results, prescription orders, and other medical instructions. The system conveys this information on screens through a series of color-coded graphics and icons, allowing nurses to tell at a glance what care a patient requires. Before the system was in place, nurses had to log on to the computer to look up lab results and check physician orders. This resulted in a time lag between when orders were entered into the system and when a nurse logged in and read them.

One of the biggest challenges in developing the system was trying to figure out what information should be displayed on the boards. The hospital collects huge amounts of data, but in many cases, that information was not readily available to the staff. This meant nurses and administrators could not see where problems were occurring and, thus, could not shift resources or move patients to resolve backups.

It sometimes took three months to mine the data to see why bottlenecks occurred. Now the staff can tell, by looking at the screens, if there is a backlog of patients waiting for X-rays. If another X-ray machine is available somewhere else in the hospital, the staff can redirect the patients to it.

The system has helped free up beds for new patients. It used to take six hours, on average, to log discharged patients out of the hospital's computer system. That meant nurses, bed controllers, or other staff would not know immediately if a patient was sent home. Secretaries on each nursing unit would walk around and check to see which patients were gone. As a result, vacant rooms often sat unused, even when other patients were waiting for them. Now, with discharge orders displayed immediately on the screens, it takes no more than six minutes to move patients out of the hospital's computer system, and to have the rooms cleaned and prepared for the next occupants.

Within six months of the pilot rollout, the system was introduced throughout the hospital except for the maternity ward, where staff members had reservations about the aesthetics of the display screens in rooms designed to evoke a feeling of hominess. But that resistance has been overcome, and the system is now being installed in the maternity ward. Outpatients who visit the hospital's diagnostic center are also being tracked with Radianse tags, which are clipped onto their clothing. This allows the staff to monitor the flow of patients and take action if any backups or bottlenecks occur.

St. Vincent's has about 140 interrogators and slightly fewer than 500 tags. The hospital has room-level coverage in several of its labs and zone-level coverage in the nursing units. Therefore, patient location is pinpointed within a range of four to six rooms.

The entire project cost an estimated $1.7 million, including the PCs, software, RFID tags, interrogators, installation, and integration, and it quickly reaped results. The number of patients discharged by noon, a key measure of operational efficiency for the hospital, climbed from about 20 to about 40 percent.

Moreover, fewer patients are being turned away for lack of beds: Patient diversions dropped by 25 percent in the critical care unit and 60 percent among medical-surgical beds. The hospital estimates that it was able to serve more patients using the RFID system, for a net revenue increase of $2.58 million during the pilot phase. And the revenue gains have continued, with the hospital taking in an additional $5.5 million between March and July 2005. The 12-month ROI for the project was 151 percent, according to the hospital.

While the reviews of the project are universally positive, there have been occasional bumps. At times, the access points required fine-tuning, with RFID readings "bleeding through" from one floor to another.

As hospitals come to understand the value of patient visibility and the relatively quick payback of RFID systems, more will embrace the technology.

## RFID Synergy at a Netherlands Hospital*

In an effort to reduce costs and improve patient safety and services, numerous hospitals and medical centers have been piloting and deploying RFID technologies to track high-value assets, patients, medical records, blood products, and beds. While many health care facilities are realizing a return on their investment, some experts believe the biggest benefits will come when an RFID infrastructure can be used to support several applications simultaneously, and when data can be combined to improve inefficiencies and automate processes.

To test for synergies among individual pilots, the Academic Medical Center (AMC) at the University of Amsterdam, in cooperation with Capgemini of the Netherlands and several hardware and software vendors, combined three RFID projects into one ambitious mega-trial: tracking and tracing medical equipment in the operating room (OR), monitoring the movements of patients and staff in and around the OR, and tracking and tracing blood products. The project received financial support from the Dutch Ministry of Health, Welfare, and Sport.

The three-pilot project was defined and developed in five stages over a five-month period, beginning in October 2005. The trial ran from July 2006 to February 2007, and was rolled out in three phases: first the patient- and staff-tracking pilot, followed a month later by the medical equipment–tracking trial, and the blood-tracking portion six weeks after that. By December 2006, all three pilots were running simultaneously.

The hospital found that it was able to save upfront costs by employing the same interrogator infrastructure for the blood- and patient-tracking pilots. In addition, it combined the patient- and blood-tracking trials to confirm that the correct patients were receiving the proper blood products. And it also combined the patient- and equipment-tracking pilots to identify which patients were treated with which medical devices.

The project scored several firsts. It was the first such project in the health care sector to have such a wide scope and document the synergy effects of running multiple RFID applications at the same time. It also documented radio frequency interference with medical devices, such as breathing machines, pacemakers, anesthesia equipment, and

---

**Source*: Rhea Wessel. (2007). "RFID Synergy at a Netherland Hospital." *RFID Journal*. Retrieved from http://www.rfidjournal.com/article/purchase/3562

heart-rate and blood-pressure monitors. Data were collected from all three pilots to use in developing a business case.

### Tracking Patients and Staff

The goal of tracking patients and staff in the operating room was to determine whether the hospital could use RFID data to optimize schedules so more patients could be treated. Each week, the pilot tracked about 20 patients having open-heart or vascular surgery. The ceilings in four AMC operating rooms were equipped with RFID activators and interrogators manufactured by Avonwood. The ceilings in the intensive care and recovery areas, as well as the hallways leading to those rooms, were also outfitted with the reader system. Altogether, about 30 interrogators were used.

Hospital staff and patients were tracked with active RFID tags. Roughly 25 people (surgeons, anesthesiologists, nurses, and orderlies) carried RFID-enabled cards they could slip into the pockets of their uniforms. Patients agreeing to be tracked were fitted with RFID-enabled ankle bracelets. The hospital's staff was much more resistant to the tags than the patients, because they did not want to feel micromanaged or be questioned about every action or movement. Some 315 patients participated in the pilot, with only two declining to participate.

Typically, a patient is moved to a pre-surgery holding area before being transferred to the anesthesia room and then on to the operating room. After the procedure, the patient is monitored in the anesthesia room before being taken to the recovery room and, if necessary, to intensive care. The study is analyzing the RFID data to determine if procedures could be changed to reduce the time a patient waits before and after the operation. This would also increase OR efficiency, enabling the hospital to treat more patients with the same resources and in the same amount of time. AMC sees little benefit to installing a permanent system to track personnel movements. The hospital further learned that it is very valuable to do the tracking, but real-time information is not always necessary to get more control over processes.

### Tracking Medical Equipment

The goal of the combined medical equipment– and patient-tracking pilot was to provide AMC with a clearer picture of which disposable products were used in which operations, and on which patients. That information could help AMC save money by controlling inventory and stock levels, and by accurately billing patients and insurers for the specific materials used.

Items required during an operation are stored in the stockroom. After the operation, used items are taken to the holding room before being discarded, and unused items are returned to the stockroom. The original idea was to employ EPC Gen 2 technology to track materials in the operating room, but interference tests, conducted with TNO, a contract research organization that works for both the public and private sector, revealed that the majority of the medical equipment was sensitive to the RFID system. The pilot was redesigned, and the system was installed in the stockroom, holding room, and hallway between the operating and holding rooms.

The pilot tracked all medical devices used in esophageal surgery, including implants, staplers, and clips. Each item, and the container in which it was stored, was tagged with an EPC Gen 2 tag from Phi Data, and patients were fitted with RFID-enabled ankle bracelets. Before an operation, the tagged container and its tagged items were interrogated in the holding room, and the unused objects were read upon being returned to the stockroom.

AMC estimates that through more efficient management, the information gleaned from this study could be used to reduce supply-related costs by about 5 percent. In addition, it says, RFID could be employed to identify devices that were recalled and to track implants, to comply with government regulations.

### Tracking Blood Products

The goal of the blood-tracking pilot was to see if RFID could be a viable alternative to bar codes for meeting European Union blood-safety guidelines. These guidelines mandate that hospitals must always know where various blood products are, and under what conditions they are being stored.

The pilot tracked blood bags used for patients undergoing open-heart or vascular surgery. Each blood bag was tracked with two tags: an EPC Class-1 Gen 2 tag from KSW and an active tag with a temperature sensor from Avonwood. Both tags, each about the size and shape of a credit card, were slipped into outside plastic pockets on the blood bag.

The tags were first read in the blood bank. An attendant swiped the passive tag across a mouse-pad-size scanning plate containing an RFID reader from Feig. The active tag was read with an interrogator mounted on the ceiling. The IDs for the blood bags' two tags were linked with the patient's ID in a computer database.

Tags were again read while en route to and inside the operating room, with the same RFID infrastructure used to track patients and staff. Before patients were given blood transfusions in the OR and intensive care unit, nurses compared the unique IDs on the patient's ankle bracelet and the blood bag to make sure they matched.

In addition to the safety benefits and the ease of complying with government regulations, the pilot showed that RFID could be used to alert attendants to potentially mismatched blood products. The system gives a much clearer overview of the actual storage locations of blood products and the time at which they are administered.

The RFID application could help AMC determine whether blood bags are stored at the correct temperature during an operation. A doctor usually orders more blood than required, to be sure enough is on hand. Following the procedure, most hospitals discard unused blood if they cannot determine whether it has been properly handled. Tagging blood bags with temperature sensor–equipped RFID tags could help reduce blood wastage by enabling hospitals to keep unused portions of open blood bags, as well as unused full blood bags.

# PART III

# Case Studies



A broad range of RFID applications are now evident. Within the supply chain spectrum, one can observe applications for improving inventory management and controlling supply chain operations. Case studies in this section are used to illustrate these broad ranges of RFID applications across industries.

Improved supply chain visibility is one of the benefits of RFID. Chapter 7: *Supply Chain Visibility* demonstrates the applications found in the retail and pharmaceutical supply chains.

Another benefit of RFID is asset visibility in the supply chain. Chapter 8: *Asset Visibility* examines how RFID is used in hospitals and health care in general, as well as the improvements in capital goods tracking.

Chapter 9: *Work-in-Progress Tracking* describes how firms are using RFID to manage their internal supply chains. Internal benefits are illustrated as firms are able to improve inventory management from the time of ordering through final inspection before shipment.

A primary potential growth area for RFID is library usage, due to decreasing library budgets and increasing demand as consumers' budgets tighten. The use of RFID to identify a library book allows the tag to be used multiple times since the same book remains in circulation a long time (i.e., it is checked in and out a number of times). Chapter 10: *Library Management System* provides examples of the benefits, such as locating missing books and identifying books that have been misplaced on the shelf.

Reusable assets are often misplaced, and the lack of visibility of their movement can lead to substantial losses for companies. Chapter 11: *Returnable Asset Tracking* illustrates how RFID asset-tracking solutions enable companies to better manage their returnable assets. This RFID application also provides information about assets due back from various trading partners, and presents information regarding the status of returnable assets against associated order numbers, improving the visibility of assets possessed by different partners across the supply chain.

# CHAPTER 7

# Supply Chain Visibility

The primary concern for companies wishing to implement supply chain visibility with RFID is the expenses associated with the process. The common question encountered in the early RFID planning phases is whether to "tag everything or nothing at all." To address this question, companies can adopt standard inferential statistics, that is, begin with tagging a representative sample of shipments and use the results to fill in the gaps. Examples of implementations reveal that a small sample is sufficient to identify the value of RFID to supply chain dynamics and supply chain velocity, and to identify bottlenecks and common exceptions.

## ✓Case 7.1: Alliance, Seeonic, and UPM Raflatac Collaborate on Item-Level Retail Display*

Alliance, the merchandising and displays division of Rock-Tenn Co. (one of North America's largest manufacturers of paperboard, containerboard, packaging, and merchandising displays), developed a hardware and service offering that would take promotions management to a new level. Partnering with Seeonic, a provider of inventory management software, and RFID tag manufacturer UPM Raflatac, Alliance developed an RFID-enabled system that would allow retailers and producers to track not only new promotional displays, but also the individual items featured on those displays. According to Seeonic, the system could also be used in conjunction with a theft-deterrence system.

The pairing of RFID with promotional displays represents one of the technology's success stories in the retail industry. Tracking the location of promotional displays using real-time visibility enabled by

*Source: Mary Catherine O'Connor. (2009). "Alliance, Seeonic, UPM Raflatac Collaborate on Item-Level Retail Display." *RFID Journal*. Retrieved from http://www.rfidjournal.com/article/view/4820.

RFID has allowed consumer packaged goods companies, such as Procter & Gamble and Kimberly-Clark, to improve the marketing of their products. Better promotions execution means better sales opportunities for retailers, such as Wal-Mart and Walgreens.

Alliance's new offering builds on the company's existing reusable product display system, known as MAXRPM, which reduces the packaging waste and shipping costs associated with single-use promotional displays. Produced with metal frames, MAXRPM displays accommodate corrugated shelving and signage swapped out with each new product promotion. The frame is continuously reused, and requires less corrugated material for each promotion, thus lowering the size and weight of the promotional elements shipped to retail stores. Working with Seeonic, Alliance is mounting a battery-powered EPC Gen 2 RFID interrogator onto the metal frame. The reader's antennas are positioned so that RFID tags attached to the products placed on the display will be interrogated when the device is powered on. The reader would then forward this data, via a Wi-Fi-enabled Internet connection, to a Web-accessible business intelligence service and relational database, SmartWatch, hosted by Seeonic.

Using SmartWatch, the manufacturer can monitor the quantity of new products on each display shelf, in real time. As a result, there is much more detailed visibility than just tagging the display and not the items it contains, which is how promotion display tracking has been performed (to date) at major retail locations. SmartWatch could be employed to automatically alert store managers and product suppliers, through e-mail or other communications, when stock levels on the displays fall too low and need to be replenished. Tomorrow's Mother (a maternity apparel company) has used Seeonic's system to track its products sold in department stores.

Additionally, the SmartWatch software can be programmed to alert store managers when multiple units of the same products are removed from a shelf on the display at the same time. This could signify that a type of theft called a sweep, in which thieves remove large numbers of the same product and toss them into a bag or otherwise conceal them, is in progress. Based on a sweep alert from SmartWatch, store managers might send security personnel to the display area, or review security camera tapes to determine whether a theft is taking place.

In 2006, Alliance acquired partial ownership of Goliath Solutions, which provides Walgreens with its promotional-display tracking system. But the Goliath system is dissimilar from what Alliance hopes to accomplish with Seeonic. Seeonic's focus is inventory control on a real-time basis, but Goliath's focus is helping retailers comply with point-of-sale display programs.

Technology alone cannot improve business operations. Even with the help of an automated system, retail store employees need to act on the data collected and ensure the physical sales-floor inventory is adequate. This was evident in Procter & Gamble's decision to halt its promotional-display tagging project, wherein contract manufacturers placed RFID tags on displays sent to RFID-enabled Wal-Mart stores. While P&G's pilot program showed that RFID had the potential to improve promotional effectiveness, it was not leading to better promotional compliance among Wal-Mart's sales associates.

Placing an RFID tag on each product placed on a display would not make good business sense for all products. Manufacturers would have a hard time justifying the added expense of tagging low-cost items, such as soap or snack foods. But one product category in which this item-level approach may be embraced is entertainment, specifically newly released movie DVDs. Because they are relatively expensive (compared with many consumer packaged goods), often the target of thieves, and in high demand for just a short time following their release, DVDs make a good candidate for item-level tagging and real-time tracking during their initial promotional period. Tesco, in fact, has tested an item-level tagging system with RFID-enabled shelves.

## Case 7.2: Gillette (2006)*

In 2006, at the time when television viewers got their first look at Gillette's new Fusion razor in a commercial aired during a Super Bowl game, Procter & Gamble's EPC team in Boston got their first virtual look at Fusion razors being brought out to the shelves of 400 retail locations thousands of miles away. RFID implementation provided this visibility for Gillette. Fusion was the first Gillette product to be completely EPC-supported at the time of launch (Fig. 7.1). RFID technology facilitated Gillette to get the product on store shelves 11 days faster than its normal turnaround time for product launches, translating into 11 days of sales in 400 stores that the retailers and Gillette might have otherwise missed. To achieve this, Gillette placed RFID smart labels on all cases and pallets of the razors shipped to the 400 RFID-enabled retail locations involved in the pilot (Fig. 7.2). Gillette also placed tags on the Fusion promotional displays it sent out to the retailers.

Visibility began as the goods arrived at the retailers' distribution centers and ended at the retailers' box-crushing machines, where reads of the Fusion case tags allowed Gillette to infer that all contents had been placed on shelves. In cases where the retailer's EPC feedback network indicated the Fusion razors or promotional displays had reached a retail store's back room, but no read events were

*Source: Mary Catherine O'Connor. (2009). "Gillette Fuses RFID with Product Launch." *RFID Journal*. Retrieved from http://www.rfidjournal.com/article/articleview/2222/1/1/

**Figure 7.1** Gillette products were among the first consumer goods with EPC support. (*Source*: http://susan.vowels.washcoll.edu/images/Future%20Store%20RFID%20Display%20Aug%202006.jpg)

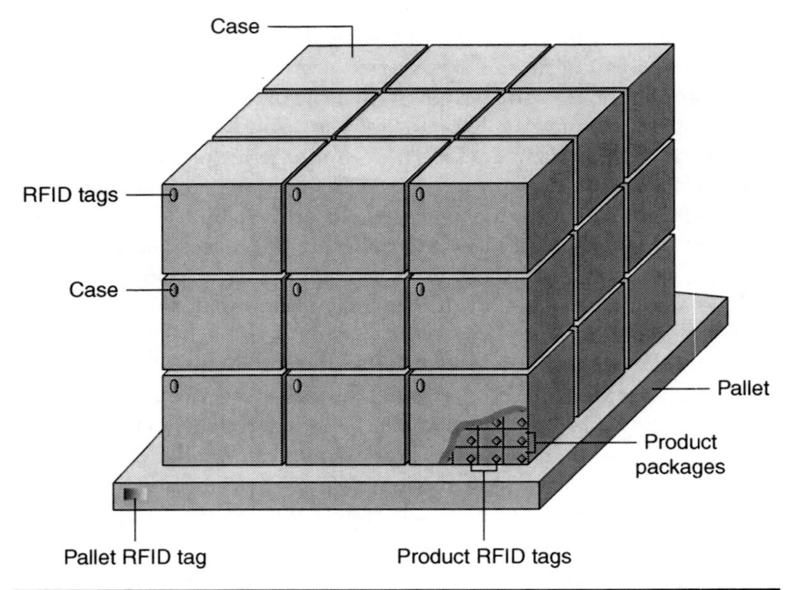


**Figure 7.2** Placement of RFID smart labels on cases and pallets. (*Source*: http://www.jefflindsay.com/gif/rfid-pallet.gif)

recorded indicating the goods being brought to the sales floor in a timely manner, Gillette contacted the managers of those stores and requested the razors and displays be brought out. Due to the high level of visibility, Gillette was able to achieve a 92 percent of its target product availability within three days of the Fusion launch. This compares with industry averages of 60 to 80 percent. And EPC-enabled stores realized greater sales than the control stores. In addition, the Fusion packaging was designed from the beginning to help RFID tags work reliably on cases and pallets. Gillette forecasts a 25 percent return on its RFID investment over the next 10 years, through increased sales and productivity savings.

## ✓Case 7.3: Charles Vögele Group*

Charles Vögele, the largest clothing retailer in Switzerland, sells fashion products at 851 stores throughout Europe, including Poland, Germany and the Czech Republic as well as Switzerland. It has 7800 employees and reported annual earnings of 1.5 billion Swiss francs ($1.3 billion). Because the retailer has a relatively complex supply chain with products manufactured in Asia and passing through several distribution centers before arriving at one of its stores, the company wanted greater visibility of its products. It also wanted to identify its "black holes," the possible problems in the supply chain. In order to improve customer service, Vögele also undertook item-level tagging (Fig. 7.3).

**Figure 7.3** Charles Vogele item-level tagging for improving visibility in the supply chain.

*Source: Claire Swedberg. (2009). "Charles Voegele Group Finds RFID Helps It Stay Competitive." *RFID Journal*. Retrieved from http://www.rfidjournal.com/article/view/4836

The retailer found that using RFID technology reduced both its stock outages (stockouts) and the time spent counting inventory by 50 percent. The company's head of supply chain recommended that despite the economic recession, the best time for an investment in RFID technology by retailers is now.

## Case 7.4: Intermountain Healthcare: Using RFID to Improve Laboratory Testing*

Intermountain Healthcare, a nonprofit integrated health care system consisting of 21 hospitals and more than 100 clinics in Utah and southeastern Idaho, is known for its innovative use of technology to improve services. Intermountain has been developing computerized health records since the 1970s, and is now working on a project with GE Healthcare to develop the next generation of electronic medical records. The health care provider serves as a strategic development partner of GE Healthcare in building advanced decision support and knowledge management tools, and other functionality for GE Healthcare's Centricity system such as integrated clinical, financial, and administrative systems.

In one of its latest ventures into technology, Intermountain Healthcare has implemented an RFID system to improve the speed and accuracy of laboratory testing. The company's lab directors and managers took note of research data from a hospital in Montreal indicating that each handoff of a lab sample from a physician to the courier to the lab, for example, added approximately 10 minutes from the time of specimen collection until the physician receives test results. So if the process involved four handoffs (which is not uncommon in a typical testing scenario), there could be about 40 minutes of additional processing time.

Intermountain decided to implement an RFID system to automate the tracking of laboratory samples. The primary drivers for the automation system were simply the needs to increase efficiency, decrease the variation in process times, and decrease staffing requirements. Company officials had considered a partial automation solution known as front-end automation, in which some of the steps in the testing process are automated and others remain manual. But they decided to automate the entire process in order to eliminate the handoffs.

Intermountain selected the Accelerator Automatic Processing System, manufactured by Inpeco in Milan, Italy, which is marketed in the United States by Abbott Laboratories. The system automatically delivers specimens to centrifuges, lab analyzers, and storage facilities. It

---

**Source*: Bob Violino. (2009). "A Tech-Savvy Medical Organization Gives the Thumbs-Up to RFID." *RFID Journal*. Retrieved from http://www.rfidjournal.com/article/view/5172.

comes with configurable middleware (called Instrument Manager), which manages and tracks the lab's data. It is designed to improve operational efficiency, reduce processing errors, and help make turnaround times on samples more consistent (i.e., reduce process variation). The Accelerator Automatic Processing System was deployed in March 2007 at Intermountain's Central Lab in Salt Lake City, a stand-alone facility located on the campus of Intermountain Medical Center, Utah's largest hospital. The Central Lab was designed to handle many of the nonemergency functions of Intermountain's seven laboratories, partly to create sufficient sample volume to utilize more efficient technology.

### How It Works

The technician registers an incoming tube containing a specimen, places a bar code label on that tube, then places the tube on a rack and puts the rack on an input bay. At this point, the Accelerator Automatic Processing System takes over. It picks up a tube and places it into a carrier sitting on a conveyor belt, with each carrier containing a passive RFID tag.

When a tube is placed in a carrier, the system "marries" the bar code identification of the tube with the RFID tag on the carrier. Afterward, the system tracks the tube's position by the position of the carrier as it passes in front of RFID interrogators. That information is sent via a wide-area network to the Instrument Manager software. This is then coordinated with Intermountain's laboratory information system software to determine what testing needs to be performed on the particular sample.

### Input-Output Station

The conveyer system delivers the tube to a centrifuge, then routes the tube to the appropriate analyzers and instruments that actually run the lab tests, such as measuring various constituents of a blood sample, required for that particular sample. Once all tests are completed, the system routes the tube to a storage module, and after a designated hold time, it is routed to a disposal unit. The system utilizes roughly 20 RFID interrogators.

The laboratory automation system has significantly improved workflow and produced lab results faster. The system has changed processes so that many routine and repetitive steps are now automated rather than done manually.

Prior to the automation, a sample tube would arrive in the lab and an individual would manually register that tube into the computer system, place a bar code label on the tube, then place the tube in a centrifuge and run the centrifuge for approximately seven minutes. The worker would then take the tube out, walk it over to one of the analyzers, and place it on a rack on a table. Another person running the analyzer would pick up the tube and place it on the analyzer, and then begin the testing.

If the sample needed to be tested on a second analyzer, then the worker would take the tube to the second analyzer and place it on a rack. From there, another person would load the tube onto the second analyzer. When the testing was completed, the tube would be placed in a rack and taken to a storage refrigerator. There it would be held for several days in case additional tests were required. After the hold period, someone would pick up and discard the tube.

Among the key benefits of the automation system are greater productivity, decreased cost per test, and more predictable process times. Additionally, the specimens can be located more quickly, and any unacceptable samples can be identified with greater speed.

When Intermountain first tested the automation system at the Central Lab, the goal was to have a 30-minute lab turnaround time for the basic blood chemistry tests. Multiple tests were conducted to simulate this requirement in order to determine how the system would perform. After several adjustments (tweaking) of the interfaces between the analyzers and the automation system, the lab is able to meet the 30-minute turnaround 80 percent of the time. The next step is to increase this performance measure to 90 percent. In addition, as volumes of lab samples continue to increase, the facility expects to be able to increase sample throughput volume up to 30 percent before needing to add more employees.

One main implementation challenge of the system was learning how to restructure workflow to realize the new technology's benefits. Some employees had to learn new roles as well as getting trained in how to use the automation system. Also, some types of samples could not go onto the automation line. Hence, Intermountain had to figure out how to keep some nonautomated processes working in parallel.

Another challenge was the interface problems that arose between the automation system and the laboratory information system at Intermountain. Likewise, the interface between the automation system and the various analyzers used in the lab posed problems. Having the various systems work together seamlessly required vendor cooperation and support. Intermountain benefited from having long-term relationships with the vendors for the analyzer, automation system, and laboratory information system.

## ✓ Case 7.5: Integris's Journey to RFID*

Hernias are painful—and not just for patients. With more than 750,000 hernia operations performed annually in the United States, hospitals must stock a large number of hernia mesh patches in various models, sizes, and styles. In most facilities, the inventory and expiration dates

*Source: Jennifer Zaino. (2008). "Integris' Journey to RFID." RFID Journal. Retrieved from http://www.rfidjournal.com/article/view/4098

of these implants are tracked manually, which can result in costly errors and lost revenue.

Nurses often pull multiple hernia meshes from a surgical cart to ensure that a doctor has the correct size on hand in the operating room (OR). It is very common for nurses to forget to return the unused patches to the cart after a procedure's completion. Equally problematic is that the expiration dates of such products are not visible in a hospital's information system. So it may not be discovered that a product is no longer viable until it is too late to return it to the manufacturer for credit. What is more problematic, due to inconsistencies in how data about hernia meshes is manually entered into an OR log, patients could be charged for a different item or not charged for items used during their surgeries.

Integris Health, the largest health care provider in Oklahoma, employed RFID technology to overhaul how the organization tracks hernia meshes, thereby reducing the likelihood of expired and missing products and bringing a hard dollar return on investment.

But the vision extended beyond tracking hernia meshes. If the pilot test were successful, it was believed it could be the beginning of an enterprise-wide RFID deployment at the provider's network of 13 health care facilities.

The pilot test proved the technology works and delivered a return on investment (ROI). Extrapolating the savings realized from tracking hernia meshes to include all products utilized in the Southwest Medical Center's OR, the ROI would be nearly $1 million. It was projected to approach $7 million were the model extended to include Integris's three metro facilities.

Integris's enterprise-wide usage of RFID within its facilities will best be accomplished by developing a shared risk and reward program in partnership with groups such as Novation, the purchasing group of the Voluntary Hospitals of America that handles supply buys for more than 400 hospitals around the country. Ideally, suppliers of implants used in hospital procedures would at some point tag their items. While this would entail an additional cost to suppliers, it would in return provide them with valuable information, such as insight into product usage, which could streamline just-in-time inventory programs. The challenge is everyone seeing an added value from the process.

Another challenge is to obtain buy-in from the many physicians who work at Integris's facilities but are not Integris employees. This is critical because the hospital cannot mandate that independent doctors wear RFID badges. But if the technology could be leveraged and integrated with existing software to enable the automatic billing of the services doctors deliver to patients directly to their offices (from the hospital's systems), it might make RFID a more attractive proposition for them.

## ✓Case 7.6: Memorial Hospital Miramar Benefits from a Real-Time Locating System*

When Memorial Hospital Miramar in Miramar, Florida, opened its doors in 2005, its goal was to provide high-quality health care, efficiently and cost effectively. The full-service acute care hospital is part of the Memorial Healthcare System that has served south Florida's residents for more than 50 years. It was designed as a virtually paperless hospital along with digital systems to support the eventual move toward electronic medical records. An RFID-based real-time locating system (RTLS) was a fundamental part of that plan.

Among the administrative challenges of operating a large hospital is tracking the movements of patients. The RTLS enables hospital personnel to track the locations of patients throughout the facility, including the management of patient flow. Provided by Versus Technology, the system, known as VISion, combines RFID and infrared (IR) technology to provide the real-time location of patients. Health care employees are freed from routine tracking tasks so they can focus more on treating patients.

Hospital officials soon discovered an even greater value in the system: VISion's ability to automate a number of other workflow processes, so the institution could improve the accuracy of its information while also providing better services to its patients. This same RTLS network is now being employed not only for patient tracking, but also to manage room turnover, dispense medications more effectively, track a variety of mobile assets, and perform other functions.

### Patient Tracking

As soon as a patient arrives at the emergency room or through admitting, he or she is provided with a Versus badge that can be clipped to clothing, suspended from a lanyard, or attached to a hospital band. The badge emits IR and RFID signals that are captured by ceiling-mounted IR and RF sensors.

That individual's location can then be tracked virtually anywhere within the facility; from the emergency room (ER) waiting area to triage and treatment, or from admitting to a patient floor to discharge. If a patient moves, the location is automatically updated in the system in real time.

The infrared portion provides the level of granularity needed to precisely locate people, even down to their hospital bed. Unlike RF waves, infrared light does not penetrate walls or ceilings. So a badge's signal stays within a room, which provides definitive location

---

*Source: Bob Violino. (2010). "Memorial Hospital Miramar Builds Benefits onto Its RTLS." *RFID Journal*. Retrieved from http://www.rfidjournal.com/article/view/7431

information that can be focused to within 12 in., assuming an 8-ft ceiling. The RF signal identifies a larger zonal area. If a badge is moved to an area lacking an IR sensor, then the RF signal will notify the system that the badge is still functioning. The RF signals are also used to send messages to the system. For example, when a room has been cleaned, a housekeeper can push the Room Ready button.

The sensors communicate the IR and RF signals to collectors that convert the location-identifying electrical signals indicating where the badge was picked up by network-ready information packets. The location data is then delivered to a concentrator that translates that information to a TCP/IP message, which is a type that can be easily understood by computer software.

A system's software suite enables alert messaging, real-time and historic reporting, and integration with a variety of health care information systems (such as security and billing). The Versus system is responsible for assigning patients to beds, time stamping and capturing patient visit milestones, and providing task notification.

The RTLS solution met the hospital's initial goals. The system provides instant identification of a patient's location in real time (Fig. 7.4). This has improved communication, expediting discharge and transfer rates, and reduced the time patients wait for rooms and employees spend searching for patients. For example, when a floor nursing supervisor receives a call from the ER that a person is being admitted to the hospital, the supervisor can then access the Versus system, determine which room is available, and inform the ER employee where to bring that patient. In addition, knowing the real-time location of patients has allowed the hospital to provide more timely services, including blood draws, medication administration, and food tray deliveries.

Supervisors can find out if a patient has been waiting an exceptionally long time to be transported back to a room after being X-rayed, and can take corrective measures. The system also automates routine tasks, like data entry, so health care workers can spend more time providing direct patient care.

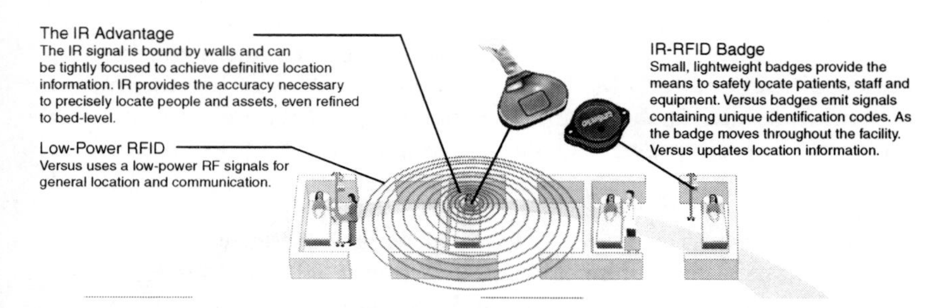


**FIGURE 7.4** How the Versus badge works. (*Source*: http://versustech.com/technology.html)

## Improving Bed Turnover

Several years after the hospital opened, administrators realized they needed to improve bed turnover rates to keep up with the volume of patients being admitted. First, they needed to get an idea of the amount of time it took to prepare a hospital room after a patient was discharged. An initial audit of bed turnover showed that the rate varied from one to two hours. A nursing supervisor had to physically walk the floors to visually check room availability, as well as discern whether the rooms were ready for incoming patients.

Hospital officials wanted to improve that rate, and determined that the Versus system could help them achieve that goal. The system identifies a room's status (clean, occupied, dirty, or cleaning in process) and alerts the housekeeping department, in real time. When a room has been vacated by a discharged patient, the system sends a message that the room is ready for cleaning. As soon as a room is clean, the system notifies admitting, the nursing supervisor, and environmental services management that the room is ready for the next patient.

According to the hospital officials, the system has improved room preparation and the average bed turnover rate has steadily improved to approximately 30 minutes. The RFID technology provides a real-time look at the bed availability throughout the facility from any computer in any department. This helps nursing management place patients in beds quicker, and with greater information accuracy, which is particularly important during a busy flu season when there could be insufficient beds and longer than average patient waiting time.

## Building on Benefits

The patient location data that is fed to the bed control system is dispersed to every department in the hospital that requires such information, including dietary, laboratory, pharmacy, and financial systems. This helps the hospital fulfill its commitment to patient safety. For example, when a patient is moved from an inpatient bed to a procedure room, or to the OR, his location is updated automatically in the system. The Pyxis system (which manages medications) is notified, and the patient's medication profile is displayed. This ensures that only the medicines ordered for that particular patient can be retrieved from the Pyxis Medstation dispensing cabinet. The system has reduced calls to the pharmacy, operating rooms, and nursing units, and has also improved communication, timeliness, and medication deliveries.

The RTLS solution was further expanded to help the hospital track the locations of millions of dollars' worth of medical and administrative equipment, such as breathing apparatus, IV pumps, and computers on wheels, carts, and digital cameras. The Versus asset tags also use dual IR-RF technology. The system enables hospital officials to monitor which devices are due for regular maintenance, and which are being underutilized. The system also helps keep carts well stocked with medical supplies, and ensures that medications are used before expiring.

Memorial Hospital Miramar is also using the system's software suite to analyze trends and determine which process improvements are necessary. The hospital is running a variety of reports related to patient throughput, such as wait and examination times, as well as the use and maintenance of hospital equipment.

## Case 7.7: Mississippi Blood Services Banks on RFID*

Few tasks pose as great a challenge for Mississippi Blood Services (MBS) as tracking inventory. The nonprofit organization collects in excess of 60,000 units of blood annually from more than 35,000 donors. At any given moment, MBS could have upwards of 2500 units of blood products on hand. And it must be ready to ship specific blood types and products, including plasma, platelets, and red cells to hospitals around the state. Blood products typically last from 5 to 42 days; frozen plasma can last more than a year.

With short shelf life, some blood products may expire in one location at the same time that other locations experience shortages of the same product, one major problem that a highly efficient inventory system could reduce. Another challenge is that accuracy in blood transfusion is a life-and-death problem, where even the slightest error can have huge consequences.

MBS supplies blood to 50 hospitals and medical facilities throughout Mississippi. Managing the inventory of blood products is a complex logistical process. For years, MBS has used bar codes and scanners to help automate processes and improve record keeping. However, in an era of growing cost pressures and increased demand for blood, the organization has recognized that RFID can help save dollars and lives.

MBS recently completed an RFID pilot that tracked 1000 bags of blood within a storage unit. Its plan over the next few years is to integrate the RFID system with its existing inventory management software and deploy the technology across the entire organization. The RFID system will provide real-time information about the location of blood, allow the organization to better anticipate shortages and distribution problems, and help improve the efficiency of the overall inventory process.

To locate plasma for a patient, for example, an employee must stand in a 15-by-15-ft or 12-by-12-ft storage freezer at –30°C (–22°F) and sort through bags. The process is just as slow and unpleasant for handling shipments of platelets or red blood cells, which are stored at a slightly higher temperature. Scanning the bar codes manually with a laser

---

*Source: Samuel Greengard. (2006). "Mississippi Blood Services Banks on RFID." *RFID Journal*. Retrieved from http://www.rfidjournal.com/article/view/2472

pen can take a couple of minutes. It is believed that RFID could simplify the validation process by streamlining a series of checks and inspections needed to ensure that the company is shipping blood to the right hospital or clinic. The technology would simplify the management of blood within the coolers.

The low storage temperature for blood, combined with its high moisture content, created challenges for developing RFID-friendly packing materials. Most systems could not overcome labeling challenges and other environmental issues, or interference problems from other electronic devices. Eventually AARFID, a firm located in Eden, New York, specializing in assembling company-specific RFID solutions and systems integration, was proposed. After several weeks of discussions, research, and brainstorming, a pilot study using 1000 units of blood products was conducted. The goal was to locate specific blood products from trays stored inside the cooler using RFID interrogators and passive tags attached to the blood bags.

The pilot study faced several challenges. First, MBS had to find a way to eliminate interference caused by the metal trays holding the blood bags that rendered reading the passive tags nearly impossible. Other electronic devices, including telephones, also caused interference problems. Eventually, after testing a variety of passive tags (ranging from 13.56 to 915 MHz), it became clear that the high water content of the blood combined with the metal trays would cause additional problems. As a result, MBS swapped the metal trays for plastic ones, moved phone locations, and replaced any metal-containing packing materials with plastic.

Another challenge was finding a label converter that could develop smart labels for the blood bags. The RFID tags had to withstand extreme temperatures, and the resins and adhesives used to attach the smart labels to the blood bags had to meet U.S. Food and Drug Administration (FDA) requirements, all while ensuring a 100 percent read rate. MPI Label Systems, based in Sebring, Ohio, devised a label utilizing an RFID inlay, human-readable text, and bar code data.

The biggest challenge was dealing with the chemical composition of the blood, which has a retuning effect on the RFID transponder. After several weeks of experimentation, as well as input from researchers at Texas Instruments (TI), it was found that the transponder would have to initiate at a frequency of 14.4 MHz and then drop to 13.56 MHz within a fraction of a second. Otherwise, the passive TI tags would not connect to the Feig Electronics interrogators at the required 20- to 21-in. range. MBS used a Zebra R2844-Z printer encoder to print the labels and verify that the tags worked correctly. Application engineers from TI and AARFID worked together to develop a unique portal enabling personnel to read the trays of blood products simultaneously.

After a worker applied the label for the specific blood product, an AARFID portal interrogator located just outside the cooler read the tag once more to verify that the system was reading the blood bags correctly as workers transported them in and out.

The system, which took about eight months to assemble, performed a data integrity check for the encoded data, consisting of the product code, FDA number, unit number, expiration date, and blood type. AARFID software controlled the printer encoder and the interrogators, managing the data that was written to each tag. It also automated the check-in and checkout of blood products from the freezers. It can perform emergency trace recall in the event that health professionals discover contaminated blood, and is equipped to handle daily and on-demand inventory of blood products, as well as order entry and order fulfillment, including packing and shipping.

Assembling the right technology was only half the story. MBS also had to change the way employees handled the blood bags. Early in the pilot test, it was discovered that an array of human factors existed. For example, workers have varying degrees of accuracy when attaching labels to the desired spot on the bag. The bar code labels have to appear in a specific area, so some workers remove the labels and reapply them, but pulling RFID labels off the blood bags could damage the transponders. So MBS has had to train workers to leave the labels in place, even if they are not in the precise position (with RFID, the exact label position is less important).

Although the RFID system works effectively and accurately, MBS faces a few additional obstacles before it can be fully implemented. All of the blood banks are regulated by the FDA as both a drug and biological service provider. Consequently, MBS must ensure that all of its manufacturing processes and systems meet FDA guidelines.

Moreover, the blood bank needs the per-unit label price to drop from about 92 cents each, as it is currently, to 25 cents apiece before it will see the initiative as cost effective. MBS believes it can get the price down to 30 cents each by buying in large volumes. In addition, it must further integrate systems with its blood bank inventory software and achieve integration with hospitals and other medical facilities across Mississippi. At present, MBS is working with other blood centers and the International Society for Blood Transfusion (ISBT) to finalize a group of standards and make RFID tags compatible across the health care industry.

MBS executives believe RFID will play an important role in managing blood supplies in the years ahead. Therefore, they are working with representatives of several leading blood centers, research institutes, and medical equipment manufacturers to develop global standards for managing blood supplies. That could hasten the adoption of RFID in the industry and boost safety and performance.

# CHAPTER 8

# Asset Visibility

Increased asset visibility is a major benefit of RFID. The case examples presented in this chapter describe applications in hospitals and health care and include capital goods tracking. These cases range in details and illustrate the many benefits of the application.

## Case 8.1: Army Medical Center Looking to Boost Asset Awareness*

Walter Reed Army Medical Center, one of the largest medical treatment facilities within the U.S. Department of Defense, has entered into a four-year contract with Awarepoint to adopt a real-time locating system (RTLS) that uses ZigBee-based RFID tags. The hospital plans to utilize the system to track 4000 pieces of equipment, according to Awarepoint, a San Diego provider of RTLS technology designed for tracking assets in health care facilities. The Washington, D.C., hospital expects the Awarepoint system will help nursing staff more quickly and easily locate equipment, allowing more time for direct patient care and reducing wait times for patients requiring transport throughout the 1.1-million-square-foot facility. Administrators will use the system to generate automated messages when an instrument is due for periodic maintenance and to locate equipment in the event of a recall.

## Case 8.2: Asset Tracking Underway at WakeMed Cary Hospital†

WakeMed Health and Hospitals, which operates numerous facilities in Raleigh and other North Carolina cities, has deployed an RFID-based system employing technology from RadarFind to track assets

*Source: Mary Catherine O'Connor. (2007). "Army Medical Center Looking to Boost Asset Awareness." *RFID Journal*. Retrieved from http://www.rfidjournal.com/article/articleview/3887/1/1/.

†Source: Claire Swedberg. (2008). "Asset Tracking Underway at WakeMed Cary Hospital." *RFID Journal*. Retrieved from http://www.rfidjournal.com/article/articleview/4056/1/1/

in its two-story, 114-bed facility in Cary, North Carolina. The system automatically monitors the location of assets, alerting the staff if an asset ends up in the wrong spot. With the installation of the system, the medical center expects to better manage assets and ensure that equipment is cleaned and serviced for reuse in a timely manner. WakeMed has also agreed to allow RadarFind to test product enhancements at the Cary facility in the future.

WakeMed Cary's nursing and clinical staff regularly spends an extensive amount of time searching for such patient-care devices as infusion pumps, stretchers, telemetry monitors, and wheelchairs. Furthermore, there is often a delay before used devices are cleaned and prepared for reuse; simply because the staff is unaware they are no longer in use.

WakeMed selected the RadarFind solution in part because it required very little time to install. Rather than having to add wiring to the patient rooms to connect the readers to the back-end system, the RadarFind interrogators plug directly into an outlet and wirelessly transmit the collected tag data.

RadarFind's active ultra-high frequency (UHF) RFID transponders are attached to assets (Fig. 8.1). The interrogators capture an RFID tag's signal, which includes a unique ID number, and transmit that data wirelessly over the 902 to 928 MHz RF band, to one of 15 collectors installed around the hospital. WakeMed Cary is not using

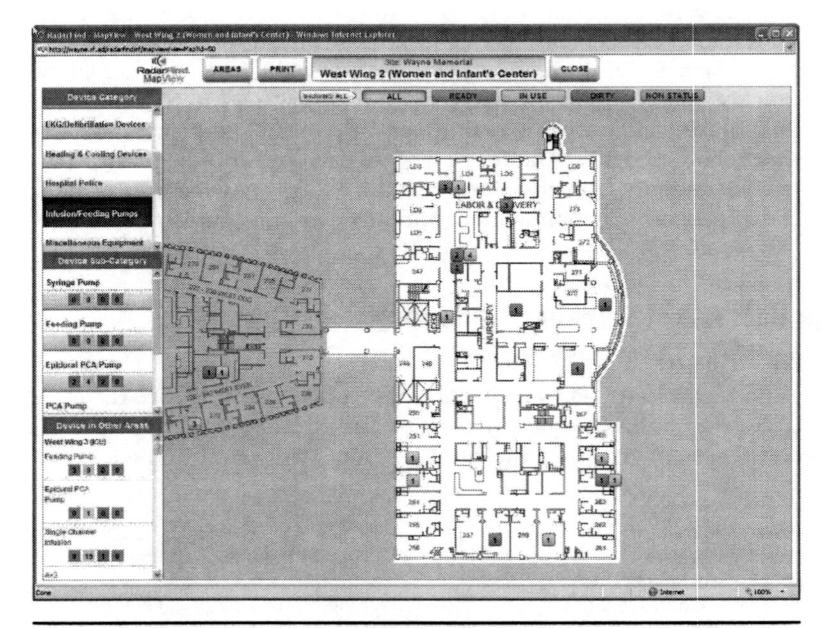


**Figure 8.1** RadarFind asset tracking displays location and status. (*Source*: Image courtesy of RadarFind Corporation.)

the readers' ability to send the collected information via the facility's power lines (employing the ANSI 709.1 protocol for power line data communications).

The interrogators calculate an item's location within several feet on the floor on which it is located, using a combination of signal strength and trade-secret technologies. In addition, the readers utilize wireless synchronous multiple-input multiple-output (MIMO) technology, a communication technique employing multiple antennas to receive data from the tags and eliminate multipath interference, thereby enabling an item's location to be determined more accurately.

The collectors then transmit the unique ID number, time, and location where they were read via an Ethernet cable to the server, located at and hosted by WakeMed but accessible by RadarFind through a virtual private network (VPN) for software updates and other services. RadarFind software translates that data and makes it available in a dashboard style, in which staff members can either type in the name of the item they are looking for, or click on a category and see a hospital map showing a dot pinpointing the item's exact location.

## Case 8.3: Carolinas HealthCare System Deploying RTLS at Its 20 Hospitals*

Carolinas HealthCare System (CHS) has implemented an RFID-enabled real-time locating system (RTLS) to track thousands of medical devices, such as infusion pumps and ventilators. The health care network, consisting of 20 hospitals in North and South Carolina, is using the RTLS to help track the devices so they can be serviced at the right time, and to ensure they are available where and when they are needed. The RTLS system leverages CHS's existing network of Cisco 802.11 Wi-Fi access points, which the organization has deployed throughout the hospitals for a variety of applications. These applications include the hospitals' computers on wheels, used to access patient information at the bedside or the point of care. The system also uses Ekahau active 2.4-GHz RFID tags, which comply with the 802.11 protocol. The tags are powered by commercial, off-the-shelf, CR2 lithium batteries with an approximate shelf life of five years.

Each tag, whenever it comes to rest following movement, emits a unique ID number. Nearby Wi-Fi access points receive tag transmissions and forward them, via a wide-area network, back to a central server located at the Carolinas Medical Center in Charlotte, North Carolina. The server runs Java-based Ekahau Positioning Engine

*Source: Beth Bacheldor. (2007). "Carolinas HealthCare System Deploying RTLS at Its 20 Hospitals." *RFID Journal*. Retrieved from http://www.rfidjournal.com/article/articleview/3704/1/1.

(EPE) software, which analyzes a number of factors, including tag signal strength, to determine an object's location (Fig. 8.2). Ekahau, based in Saratoga, California, with offices in Virginia, Finland, and Hong Kong, has designed the EPE software to allow the system to track the real-time location of more than 10,000 objects on one server, and calculate up to 600 locations per second to within an average of 1 m (3½ ft).

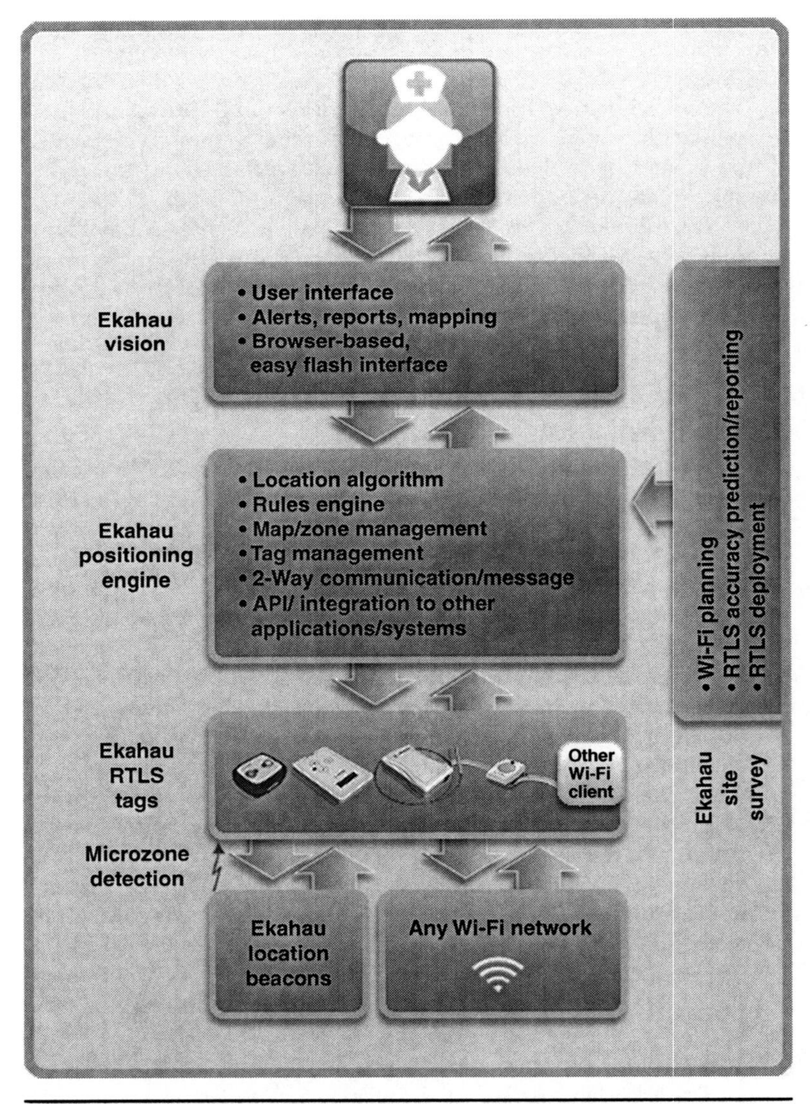


**Figure 8.2** Ekahau RTLS infrastructure. (*Source*: http://www.ekahau.com/products/real-time-location-system/overview.html)

Each tag's unique ID number has been associated with information related to that specific device, such as its maintenance schedule and location, and housed on the central server. CHS employees with appropriate credentials can use a Web browser to access a real-time view of a tracked area, as well as the location of tagged devices in that area, or search for particular devices and their related information. For example, a technician can query the system using a serial number or other descriptor to locate a device due for maintenance.

CHS will track about 5000 medical devices, most of which are required to go through preventive maintenance. The RTLS gives CHS the visibility to properly locate and maintain them. It also enables CHS to keep the most appropriate inventory on hand to meet the needs of its hospitals, and saves time for doctors, nurses, and other care providers.

## Case 8.4: AeroScout Unveils New Asset-Tracking Platform*

AeroScout, a Redwood City, California, provider of a Wi-Fi-enabled RFID real-time location system (RTLS), has collaborated with Time Domain Corp. and Reva Systems to create a unified platform that organizations can use to work multiple RFID technologies, thereby achieving more comprehensive tracking of assets, processes, and personnel.

AeroScout's advanced software, MobileView 4.0, allows organizations to utilize a single platform to access and act on data not only from Wi-Fi RFID tags but also from passive and ultra-wideband (UWB) tags (Fig. 8.3). This upgrade from MobileView 3.1 provides users with an entire organizational view to specific individual zones and even rooms, via graphical map displays, as well as numerous reports that can blend data to present historical trends and other information. In addition, users can create more types of alerts using the previous version of MobileView, and employ a range of media, including e-mail and cell phones, to send such alerts. MobileView 4.0 also acts as a middleware platform that brings together information from multiple sources and shares that data with third-party applications.

This application software launch has been attributed to the growing needs of the customer organizations to have a unified view of asset visibility within their organizations. For example, a hospital may want to use Wi-Fi-based RTLS to track the location of infusion pumps or other hospital equipment because the facility already has a Wi-Fi network. Tracking the infusion pumps makes it easier for

*Source: Beth Bacheldor. (2008). "AeroScout Unveils New Asset-Tracking Platform." *RFID Journal*. Retrieved from http://www.rfidjournal.com/article/articleview/3887/1/1/

**FIGURE 8.3** AeroScout RFID tag. (*Source*: http://medicalconnectivity.com/gems/Blog%20Photos/AeroScout-tag.jpg)

nurses to locate the devices, and by using a Wi-Fi RTLS, the hospital does not have to install a new network infrastructure. Wi-Fi-enabled active RFID tags such as AeroScout's can transmit 2.4-GHz signals and communicate their unique ID numbers to the hospital's Wi-Fi network, which can consist of Cisco access points. To compute location, the hospital can use either the AeroScout Engine or a Cisco 2710 Wireless Location Appliance.

A hospital using a Wi-Fi-based system may decide that in certain scenarios, it also wants to use UWB active RFID tags and interrogators so it can determine the exact location of assets. It may want to use UWB to document that an infusion pump was in a specific room at a given time, for instance, and within inches from a patient (also identified by means of a UWB tag), thereby indicating the pump is being used on that person.

## Case 8.5: Denver Health Adopting a Hospital-Wide RTLS System*

Denver Health has been expanding an RFID-based real-time locating system (RTLS) to track vital medical equipment at its Women's and Children's Pavilion, a facility that the 500-bed teaching hospital opened in August 2006. RTLS is expected to cover the entire hospital, an area encompassing 1.5 million square ft. The hospital is

**Source*: Beth Bacheldor. (2007). "Denver Health Adopting a Hospital-Wide RTLS System." *RFID Journal*. Retrieved from http://www.rfidjournal.com/article/articleview/3718/1/1/.

**Figure 8.4** Wi-Fi-based active RFID tag. (*Source*: http://www.engadget.com/2007/02/02/pango-unveils-wifi-based-active-rfid-tag/)

employing InnerWireless's Wi-Fi-based RTLS technology, which the company acquired when it merged with PanGo Networks in March (Fig. 8.4). Now known as Vision, the RTLS solution is a Wi-Fi-based system that incorporates VisionOS Platform middleware, along with Vision V3 2.4-GHz active RFID tags that comply with the 802.11b and 802.11g Wi-Fi standards. The middleware aggregates the location data and unique ID numbers culled from the RFID tags, as well as the hospital's Wi-Fi access points, and passes that information on to the asset-tracking software, which the hospital can then access to obtain real-time asset visibility, alerts, and reports (Fig. 8.5).

Denver Health originally opted to use a Wi-Fi-based RTLS largely because it wanted to leverage an existing Cisco Unified Wireless Network with 300 access points. The system's track record, however, is the driving force behind the expansion.

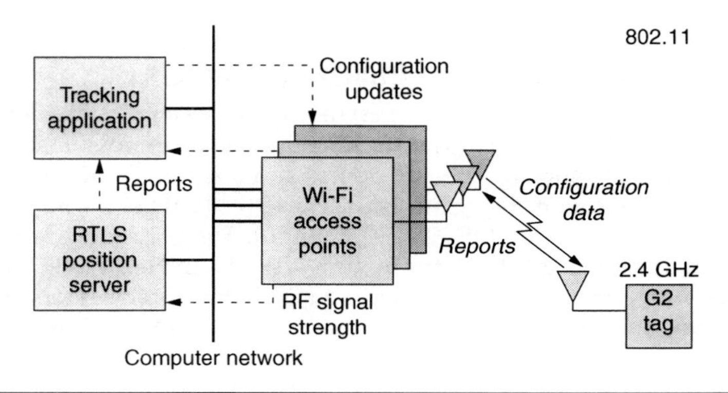


**Figure 8.5** Active RFID tracking system. (*Source*: http://www.g2microsystems.com/images/g2_modes1.jpg)

At present, the hospital is in the midst of increasing its supply of medical devices, such as infusion pumps, wheelchairs, and wound vacuums, from just several hundred pieces of equipment to 2700. Knowing the real-time location of specific equipment has enabled Denver Health to improve its workflow. Clinicians and staff can now focus on attending to a patient's needs as compared to spending time searching for misplaced items or looking for a piece of equipment that needs maintenance. To keep tabs on maintenance schedules, the company has integrated Four Rivers Software Systems' maintenance management software with the VisionOS Platform, enabling automated alerts for maintenance schedules. The data that the Vision OS software provides is then used to streamline work process and control purchasing.

Using reporting tools within the VisionOS Platform, Denver Health can analyze operations and make any necessary changes to processes and equipment inventory. For example, after investigating data culled from tracking wheelchairs, the hospital discovered that many unused wheelchairs were spread throughout various departments. By knowing the chairs' locations at any given time, the hospital has been able to cut its wheelchair inventory.

In another example, the hospital has been able to reduce inventory of wound vacuums, used to drain fluids from patients' wounds. When a patient is admitted to the hospital and treated with a wound vacuum, the device is left in the room where that person was first treated. A few days later, the patient might be scheduled for surgery, at which time a wound vacuum would again be needed. Rather than retrieving the device from the first treatment room, the OR surgeon might lease another wound vacuum for the same patient. With Vision, double-leasing has been reduced, which helps avoid unbillable costs. The hospital can use the software to automatically search for and locate the original wound vacuum leased for a particular patient, rather than having to lease two of the devices for the same person.

## Case 8.6: Emory Healthcare Tracks Its Pumps*

Emory Healthcare, Georgia's largest health care system, has deployed GE's IntelliMotion RFID asset-tracking system to improve management and utilization of infusion pumps and other high-value equipment. IntelliMotion comprises 2.45-GHz active RFID tags compliant with the newly ratified ISO 24730 standard. The platform is manufactured by the RTLS platform provider WhereNet, along with WhereLAN location sensors, which act as tag interrogators and locators, and WherePort exciters, which serve to wake up WhereNet

*Source: Mary Catherine O'Connor. (2007). "Emory Healthcare Tracks Its Pumps." *RFID Journal*. Retrieved from http://www.rfidjournal.com/article/articleview/3311/1/1/.

tags in a dormant, energy-saving mode. GE sells the WhereNet hardware under its IntelliMotion brand, coupled with GE's IntelliMotion Web-based asset-tracking software.

Emory University Hospital, Emory Crawford Long Hospital, and Wesley Woods Geriatric Hospital each have 800 infusion pumps that have already been tagged and are presently being tracked. Before using RFID to track the items, the hospitals performed regular manual inventory counts that told them only how many used pumps were in the soiled-utility areas of each hospital, and how many cleaned pumps were in stock and ready for issuing to patients. What they did not know was how many pumps were being used, how many had been used but had not yet been brought to the soiled-utility area, or how many might have been removed from the facilities.

This information was inadequate for hospital staff to feel confident they had all the infusion pumps they might need. As a result, inventory managers often rented additional infusion pumps to supplement their in-house stocks, so they would be ready to respond to nurses' requests for the pumps. Under the new system, RFID interrogators installed at the doorways to the soiled-utility rooms record when tagged pumps enter or leave the rooms. IntelliMotion pulls this data into inventory lists detailing how many pumps are in the utility areas at any given time.

Location sensors are also installed in the entrance and exit points of the inventory areas, providing an inventory of tagged pumps available for use. This helps keep the hospitals' inventory records more accurate and more quickly updated. The hospitals now have a level of asset visibility they never had before. Location sensors installed at choke points and other strategic locations across the hospital wards read the WhereNet tags as the pumps are brought into service in those wards. Location sensors also read the tags as the pumps are removed from those areas. Armed with this new data, the hospital team can cross-reference lists of infusion pumps and registered patients in each ward.

Occasionally, ambulance personnel tranferring a patient to another facility take an infusion pump used to support the patient. These technicians are supposed to bring their own infusion pumps for use while the patient is in transit, but they sometimes forget to do so and often fail to return the infusion pumps. WhereNet location sensors have now been installed at the hospitals' exits, so when someone takes an infusion pump, the system can quickly remove it from available inventory. Later system development will enable these sensors to trigger alerts sent to pagers, enabling hospital staff to know as soon as an infusion pump leaves the premises. The alerts will remind the team to check which ambulance services were removing patients at the time the location sensors read the pump's tag. Later, they can call the ambulance company and request the device be returned to the hospital. The IntelliMotion software updates the inventory lists each time tagged devices are brought into or out of the doorways or other choke points where location sensors have been installed.

## Case 8.7: German Researchers to Test Networking Tags for Assets, Blood*

At Erlangen University Hospital, Erlangen, Germany, a consortium that includes the Fraunhofer Institute for Integrated Circuits (Fraunhofer IIS) will soon carry out a six-month trial of a wireless sensor network developed to locate medical equipment and monitor the "cold chain" for blood, a temperature-controlled supply chain (Fig. 8.6). Fraunhofer IIS's communication networks department developed the system as part of the OPAL-Health project, which was partly sponsored by the German Federal Ministry of Economics and Technology's SimoBIT program. SimoBIT supports the development of secure mobile information technology (IT) applications for midsize companies and public agencies.

The idea for the project first emerged during discussions held in 2006 between Fraunhofer IIS and industrial partner Delta T, which operates in the blood-transport sector. During the mid-2006, all five project partners met to identify numerous manual processes in blood transport, as well as the tracking and locating of expensive medical

**Figure 8.6** Monitoring the cold chain for bags of blood. (*Source*: http://www.german-info.com/images/edu_images/Intelligent-blood-bags-large.jpg)

*Source*: Rhea Wessel. (2009). "German Researchers to Test Networking Tags for Assets, Blood." *RFID Journal*. Retrieved from http://www.rfidjournal.com/article/view/5143.

equipment that is often the property of a single hospital division but shared by many. The partners designed the system to reduce manual tracking processes for the machines and to improve the efficiency of those processes.

For example, if a patient is attached to a portable heart-monitoring device after surgery, that device (which is the property of the surgical division) is loaned out to the section of the hospital that operates the recovery room when a patient is moved from one area to another. The surgical division, which must account for its expensive medical equipment, runs the risk that the heart-monitoring device might become misplaced within the hospital, or be stolen. To prevent either scenario from occurring, OPAL-Health's partners sought to monitor the devices remotely. They considered passive RFID for the job, but ruled out that option since passive RFID tags are read only when excited by an interrogator, and the partners wanted constant monitoring. In addition, consortium members were concerned about potential interference that RFID transmissions may cause for sensitive medical equipment.

Another idea was to employ Wi-Fi tags in conjunction with WLAN technology. However, few hospitals in Germany are completely outfitted with a Wi-Fi network. Ultimately, the partners ruled out Wi-Fi tags because they are too large and, due to WLAN standards, use substantial energy, thus leading to a considerably shorter operating life for the tags. OPAL-Health's partners opted to build a network of wireless sensors it calls smart objects (Fig. 8.7). These tags are battery-operated and contain a microcontroller, a wireless transceiver, software, and temperature sensors. The tags form a network, with one tag communicating wirelessly with another. The wireless

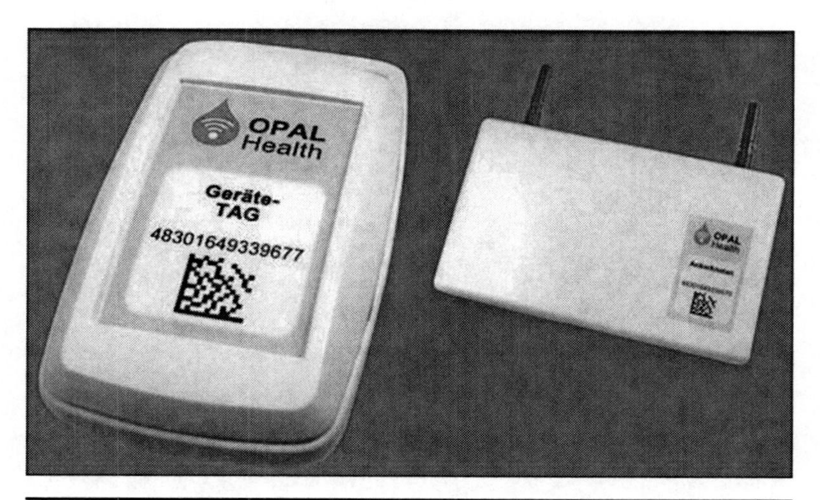


**Figure 8.7** Wireless tag and anchor node in OPAL-Health system. (*Source*: http://www.rfidjournal.com/article/print/5143)

tags autonomously determine their position within the sensor network, based on the received signal strength of fixed anchor nodes. The tags utilize Fraunhofer IIS's patented media-access protocol. The *Slotted MAC* (as it is known) allows for energy-efficient and secure data communication. The tag's unique ID number and information on temperature and position are passed along the network until reaching the hospital's clinical information system, which then uses the data to map the locations of tags and alert users if temperatures move outside of the desired, preset range.

## Case 8.8: Howard Memorial Finds RFID Keeps Assets from Getting Lost*

Frank R. Howard Memorial Hospital, a nine-building facility in Willits, California, has begun employing a ZigBee-based system to help track location of items, as well as tracking when they require repair or cleaning, and when they are being removed from the buildings. The system was provided by medical equipment solutions firm Skytron, with Awarepoint hardware complying with the IEEE 802.15.4 standard (Fig. 8.8). The system was installed in the spring of 2009, and in its first months it reduced the incidence of lost items from up to 30 per month down to zero. Some of the nine buildings at Memorial

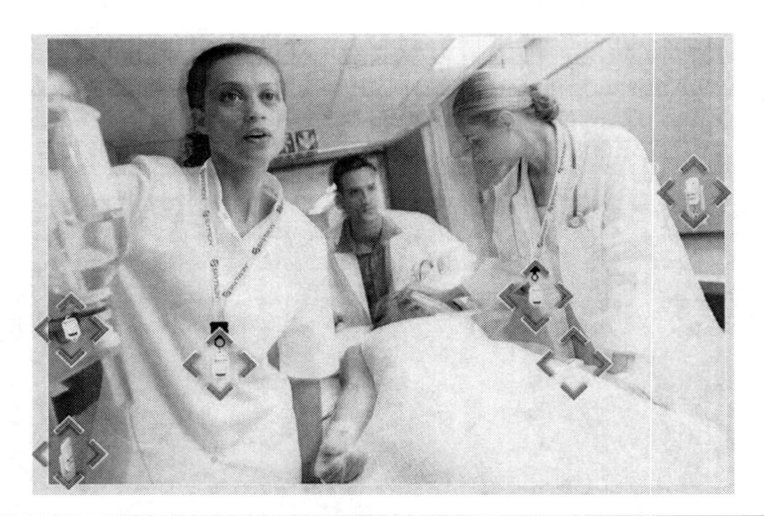

**FIGURE 8.8** The tags are used to determine position within the sensor network. (*Source*: http://www.skytron.us/s-asst-mgr.htm)

**Source*: Claire Swedberg. (2009). "Howard Memorial Finds RFID Keeps Assets From Getting Lost." *RFID Journal*. Retrieved from http://www.rfidjournal.com/article/view/5244.

Hospital are attached via hallways, while others stand alone. The rural hospital needed a system that would help its biomedical department and other staff members gain immediate information regarding which building an item is in (and in which location within that building) as well as the maintenance and cleaning status of any specific assets, such as wheelchairs, wheeled workstations, portable thermometers, and X-ray image intensifiers.

Inefficiencies were experienced in locating equipment when it was time for servicing assets. After seeking solutions, the hospital chose Skytron's ZigBee-based system in part because the company was already a contracted vendor for the facility's health care alliance, Premier. The challenge for Skytron was to complete the project in about two weeks, because Howard Memorial's CEO intended to showcase the solution at an Adventist Health West conference for hospital CFOs and CEOs.

Several challenges related to the installation were experienced. The hospital needed the system to operate within all nine buildings. Because assets were often moved from one building to another for servicing, they would occasionally leave the ZigBee coverage area. Consequently, the system needed to allow for that activity, while also being able to recognize any unusual asset movements that could indicate a theft was under way.

Skytron installed approximately 90 ZigBee sensor units, which function similarly to RFID interrogators, throughout the facility's nine buildings, and the hospital attached about 300 Awarepoint battery-powered tags to the assets. Each tag stores a unique ID number that it transmits to those sensor units. It also comes with a slide switch that workers can move from one side of the tag to another, thereby changing the transmission data to indicate a change in the asset's condition, such as needing to be serviced.

With the system in place, a tag is usually in read range of at least 15 sensor units within the vicinity to help pinpoint its location which, in turn, transmit to a bridge unit. The bridge unit calculates the tag's location within several meters, then sends that information, along with the tag's ID number and status (such as requiring service), via a wired LAN connection to the hospital server, which uploads it to Skytron's network operations center (Fig. 8.9).

Skytron software then interprets and displays the data on a Web site accessible only by authorized users. Employees must sign onto the system using their user name and password, and are granted access to predetermined amounts of data regarding the asset's location and status. Staff members can utilize the system to locate an item, determine its status and history, and learn when that asset may need servicing in the future.

The system can also send alerts to authorized users when an unexpected action occurs, such as a specific item leaving at an unscheduled

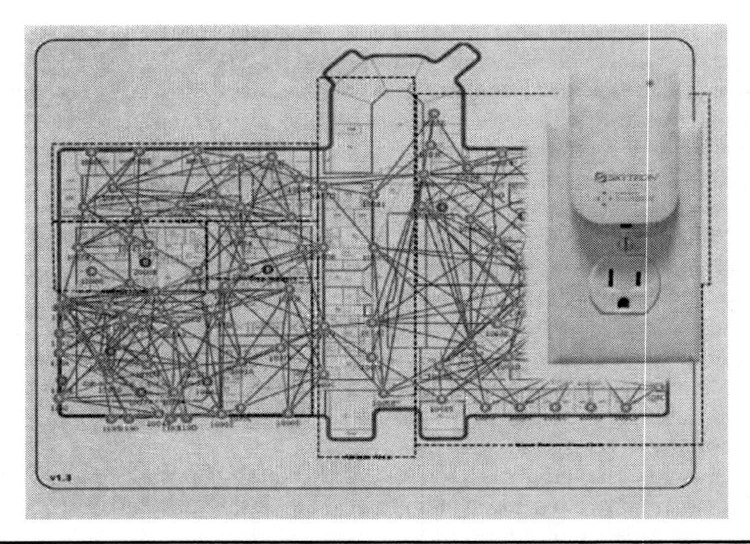

**Figure 8.9** Skytron's network used to display assets and movement. (*Source*: http://www.skytron.us/s-asst-mgr.htm)

time or through an unexpected doorway, thus indicating a possible theft. An integrated optical sensor can detect if a tag has been removed from the item to which it is attached, causing the tag to send an alert. A built-in accelerometer then helps the system determine whether that tag is moving or stationary.

The system is also self-healing and self-calibrating, so that if one sensor node goes down, data is still transmitted through the others. Skytron software monitors the nodes' health, and sends an alert in the event that one of the sensors is not operating properly. Currently, the hospital's staff uses the Skytron Asset Manager software to search for items an average of 750 times per month.

## Case 8.9: Jackson Memorial Enlists Thousands of RFID Tags to Track Assets*

In January 2009, Jackson Memorial Hospital admitted eight critical patients simultaneously: all American cruise ship passengers injured in a bus accident in Dominica and airlifted to the Miami hospital. Doctors wanted to be sure they had all of the proper equipment available to treat the injured. And they did by using a newly installed real-time locating system (RTLS) from Awarepoint that the hospital is

*Source*: Beth Bacheldor. (2009). "Jackson Memorial Enlists Thousands of RFID Tags to Track Assets." *RFID Journal*. Retrieved from http://www.rfidjournal.com/article/view/4638.

utilizing to track thousands of items throughout its nearly 4-million-square-foot, multi-building campus. Using the RTLS, the head of respiratory care was able to find a ventilator in an area it should not have been in, and where staff never would have looked. The RTLS allowed the hospital to have this extra ventilator ready when the patients came in.

Jackson Memorial Hospital, a 1500-bed teaching hospital that is part of the Jackson Health System, affixed active RFID tags on everything from infusion pumps to wheelchairs to ultrasound machines. The hospital aimed to tag a total of 12,000 assets as implementation rolled out, affording everyone from doctors and nurses to pharmacy personnel and therapists the ability to search for and locate equipment throughout the facility.

When tags are affixed to the items, each asset tag's unique ID number is correlated with a visible asset control number on a bar code already affixed by the hospital. The tags operate at 2.48 GHz, transmitting their unique ID numbers over the 802.15.4 (ZigBee) communications protocol, to small receivers (which Awarepoint refers to as sensors) that plug directly into standard 120-V AC wall outlets. A tag or sensor can pass data to a main access point (known as a bridge) by first transmitting it to another tag, which then forwards the information to a third tag or sensor, or to the main access point, depending on whether the second tag is in range of the main access point. In the Awarepoint network, a tag can send data to a bridge through up to five other tags and receivers. The receiver forwards a tag's ID number and signal strength to a bridge, along with its own ID number and the time it read the tag, as well as the ID of the transceiver that may have previously picked up the tag's signal. The bridges link, via an Ethernet cable, to a central Awarepoint server that calculates the locations of all tagged assets and then displays that information on a map of the facility. Any computer linked to the system's local area network (LAN) can access the map and employ Awarepoint's software to search for a specified type of asset.

According to the company, the software can provide the item's location to an accuracy level of 1 to 3 m (3 to 10 ft). The Jackson Memorial deployment, the company reports, provides an average location accuracy of 1.5 m (4.9 ft) or better throughout the hospital, including in-room locations, as well as hallways and other defined areas. Awarepoint utilizes a proprietary algorithm to determine asset locations, based on the tags' RF signal strength. Generally, the firm reports, Awarepoint's RTLS requires one sensor per 1000 square ft, and one bridge per 20,000 square ft.

Jackson Memorial Hospital's employees can use the real-time locating system at any of the health care organization's networked computers. By logging into the hospital's intranet and clicking on an icon representing the RTLS, the staff can perform searches for medical

**Figure 8.10** Awarepoint system can be used to track assets. (*Source*: http://www.awarepoint.com/patient-tracking-patient-flow.shtml)

equipment in much the same way searches are conducted using Google. Workers can search for "wheelchair" to locate all of the facility's wheelchairs, for instance, or they can narrow the search to a particular area or floor (Fig. 8.10). Searches can also be performed using an item's specific asset-control number.

In addition to the asset tags used to locate items, the implementation includes 250 Awarepoint T2T temperature-monitoring tags, which can wirelessly monitor and maintain logs regarding temperature-sensitive assets. Jackson Memorial utilizes the tags to monitor conditions within refrigerators used to store pharmaceuticals. The RTLS is programmed to send e-mail alerts to personnel if any of the refrigerators' tags log temperatures outside of a preset, acceptable range. The hospital is also employing the T2T temperature tags to monitor its radiology data centers. Temperature tags are placed in front of server panels to monitor the ambient temperature and enable the staff to utilize external air conditioners, when required.

Awarepoint began working on the Jackson Memorial implementation in late October 2008, and the real-time locating system went live in mid-December. The hospital continues to consider opportunities to utilize the RTLS to track assets; since it began using the system, it has discovered that the technology will be useful in helping to protect more than just medical equipment. The hospital has requested an additional 400 tags to keep tabs on the televisions.

The Jackson Health System plans to expand the RTLS to other facilities as well. The current implementation covers 91 floors and 17 buildings, and includes Jackson Memorial Hospital's South, North, and West wings, the East Tower, Holtz Children's Hospital, the Ryder Trauma Center, the Mental Health Hospital, and the Highland Professional Building, as well as parking pavilions and other annex buildings. A second phase, to include another 8000 tags, is planned to involve Jackson South Community Hospital, a 199-bed acute care hospital located in south Miami-Dade County, and Jackson North Medical Center, a 382-bed acute care center located in North Miami Beach.

## Case 8.10: North Carolina Hospital Looks to RadarFind to Improve Asset Visibility*

Southeastern Regional Medical Center (SRMC) in Lumberton, North Carolina, has contracted RadarFind to install an asset-tracking system that uses active RFID tags and interrogators operating in the 902 to 928 MHz range. The tags communicate with readers that plug into standard AC outlets and have a design that keeps both outlet sockets available for use by other devices.

The hospital hopes the system can help reduce its spending. The hospital currently relies on employees taking periodic manual inventory of important devices such as wheelchairs and infusion pumps. But that system takes too much time and produces erroneous data, since many assets cannot be easily located and, thus, might not be counted. As a result, superfluous replacement equipment might be ordered. The company identifies its need to have timelier, accurate assessments of inventory levels that would minimize the challenge of performing inventory and reduce equipment costs. But another important goal in deploying the system is to minimize the frustrations that care providers, clinical technicians, and other staff members feel when they are unable to quickly locate a piece of equipment. In 2006, Wayne Memorial Hospital, a 316-bed facility in Goldsboro, North Carolina, deployed RadarFind's system and was able to save more than $300,000 in equipment expenses within a year of installation.

RadarFind's active ultra-high frequency (UHF) RFID transponders use multiple-input, multiple-output (MIMO) communication. MIMO is a wireless communication technique utilizing multiple analog signal paths among multiple antennas to transmit and receive data. The interrogators' range can be set from 3 to 150 ft. Once they receive tag data, the readers pass that information to devices that

---

**Source*: Mary Catherine O'Connor. (2008). "N.C. Hospital Looks to RadarFind to Improve Asset Visibility." *RFID Journal*. Retrieved from http://www.rfidjournal.com/article/articleview/3878/1/1/.

RadarFind calls collectors. Typically, one collector is installed on each floor of a facility. The readers can communicate with the collectors either by transmitting data wirelessly over the 902 to 928 MHz RF band, or by sending information across the power wiring. The collectors then pass the data, via a local area network, to a RadarFind server.

Each RadarFind asset tag is encoded with a unique ID number, transmitted by the tag, and also features a switch to indicate the asset's condition. When the asset is clean and available for use, a nurse slides a plastic lever to expose a green sticker. This also causes the tag to modulate a signal to denote that the asset is ready for use, so that personnel using the RadarFind software can view a floor plan of the facility and observe both the asset's location and status. Once a nurse begins using the asset, the switch is positioned to reveal a yellow sticker. This indicates to staff passing by, or to those viewing the floor plan through the software, that the asset has been assigned to a patient and is not otherwise available. Once the asset's use is complete, the nurse moves the lever to reveal a red sticker, indicating the asset should be picked up and moved to a cleaning facility within the hospital. Hospitals can use the status button to help ensure devices are cleaned before each use, preventing the spread of infection.

A specially designed asset tag for wheelchairs comes with electric-field sensing capabilities able to detect if a patient is in a wheelchair. RadarFind executives claim the tags have a battery life span of approximately six years. The batteries are not replaceable, though a RadarFind spokesperson says hospitals will likely replace many tagged assets before a tag's battery life ends.

The hospital would like to track another 1000 pieces of mobile equipment—items such as handheld radios used for communications—but the RadarFind tag's size (2.5 in. long by 1.25 in. wide) and $40 price tag make such a use impractical. However, RadarFind is developing a smaller tag that should cost less and work for tracking smaller assets. The tag would also have signal modulation to denote that the asset is ready for use, so that personnel using the RadarFind software can view a floor plan of the facility to observe both the asset's location and its status.

## Case 8.11: New York Medical Center Tracks OR Equipment for Trauma Care*

University Hospital, part of the State University of New York (SUNY) Upstate Medical University, is employing a Wi-Fi-based RFID system provided by AeroScout to locate equipment in the emergency operative

*Source: Mary Catherine O'Connor. (2008). "N.C. Hospital Looks to RadarFind to Improve Asset Visibility." *RFID Journal*. Retrieved from http://www.rfidjournal.com/article/articleview/3878/1/1/.

centers. The hospital, a Level 1 trauma center for central New York State, often has a critical need for fast medical services in its surgical rooms. The facility's operating rooms (ORs) are located in two sections of the hospital, with four surgical rooms for children on the third floor and another twelve for adults on the fifth floor. Equipment is shared between the two areas. Because of the high level of activity and limited space, the hospital also stores equipment, such as specialized operating tables and diagnostic machines, in basement storage units.

With multiple storage and operating locations, it is easy to understand how complex and how important it is to have items where they need to be for scheduled procedures, and also for unexpected needs. Before installing the AeroScout system, staff members often had to physically search for required equipment to meet unexpected and urgent needs. Surgeries sometimes had to wait until the proper equipment could be located. Quick location of this equipment was an urgent need for the hospital's operation.

However, it was not the OR equipment-tracking concerns that first drew the hospital to a Wi-Fi-based tracking system. In fact, SUNY initially sought a solution to track other equipment, particularly IV pumps. Because the hospital had installed a Cisco Wi-Fi network and access points for communications throughout its 1.3 million-square-foot facility in 2007, it wanted to leverage that technology to improve its asset visibility and ultimately, its patient care.

The hospital management decided to implement AeroScout's Wi-Fi RFID Asset Management solution, which includes the company's MobileView software to view the location of an item on a PC, as well as Wi-Fi tags that transmit to the existing Wi-Fi nodes deployed throughout the facility.

The first phase of the system was deployed in the 366-bed hospital in November 2008. When the facility's management selected the AeroScout solution, they wanted to improve IV pump utilization because the company suspected it was too low, and that with better information they could improve utilization and decrease further capital purchases. However, after the hospital tagged and began tracking 600 IV pumps, the problem of storing and locating the OR equipment brought the department's needs to the forefront. It became clear that the immediate challenge was the management of the OR equipment.

With the RFID system, employees can log into the MobileView software at the nearest PC and view a graphical interface that displays the location of an item that a staff member keys into the system, such as a specialized operating table. The department has employed a clinical engineer to manage the inspection and maintenance of hospital equipment, and tagging has been made part of this process. Thus far, the hospital is attaching tags to 3000 assets. SUNY's Central

Equipment Services department, responsible for locating items as well as cleaning and maintaining them, is tagging equipment as it arrives in that area for servicing. The hospital is continuing to tag items as they are serviced.

In addition, the facility has installed a temperature-monitoring system in approximately 100 refrigerators that store pharmaceuticals, vaccines, and bone and tissue samples. Before installing the AeroScout system in the refrigerators, nursing and pharmacy workers were tasked with manually tracking temperatures in the refrigerators several times each day, then recording those measurements on paper. The system was slow and had the potential for errors.

With the AeroScout system, Wi-Fi tags attached to refrigerators transmit temperature data at preset intervals. This information, as well as the date and time, is transmitted along with the unique ID number of the tag within the refrigeration unit. The MobileView software associates the temperature data with a specific refrigerator. The hospital can record data regarding refrigerator health, as well receive alerts if the temperature reaches an unacceptable level, which could put any temperature-sensitive contents at risk. With greater productivity, equipment utilization, and OR throughput, the company anticipates a payback period of less than a year.

## Case 8.12: Philips Introduces Asset-Tracking System for Health Care*

In 2006, Royal Philips Electronics' medical-systems division introduced a new asset-tracking solution leveraging RFID and standard Wi-Fi-based wireless networks. This new division will help hospitals not only track medical devices, but also monitor how staff members handle such devices. Built on a suite of software Philips designed to help health care institutions manage their staff, patient care, assets, and operations, the system has been integrated with real-time locating system (RTLS) hardware and software from RTLS specialist AeroScout.

The RTLS system includes AeroScout T2 active RFID tags; AeroScout Exciters, which activate the tags to transmit their ID numbers, thus providing location information whenever tags pass by their locations; and AeroScout Engine, a software component that calculates tag locations by processing data from Wi-Fi access points and the active tags. AeroScout's MobileView software associates each tag's ID with the corresponding device, while also collecting and storing location data. The MobileView software can also portray location information on a map, or in a table or report format.

---

*Source: Mary Catherine O'Connor. (2008). "N.C. Hospital Looks to RadarFind to Improve Asset Visibility." *RFID Journal*. Retrieved from http://www.rfidjournal.com/article/articleview/3878/1/1/.

Hospitals can apply T2 tags to a variety of medical equipment, such as ultrasound machines, electrocardiogram (EKG) machines, and compression pumps used to treat thrombosis and improve circulation. Each tag transmits a 2.4-GHz signal carrying the tag's unique ID. The transmission uses the IEEE 802.11 air interface protocol, known as Wi-Fi, so organizations with existing networks of Wi-Fi access points need not install proprietary interrogators to collect tag data. More and more hospitals are installing Wi-Fi networks. The combination of Philips's medical management software and AeroScout's Wi-Fi-based RTLS system provides hospitals an asset-tracking solution that is less expensive and easier to install. This would help to make the return on investment more obtainable, more quickly.

Philips will help deploy the asset-tracking system, tailoring it to each hospital's specific needs to help the organization solve problems with productivity, regulatory requirements, utilization, theft, and loss. With its health care expertise and an in-depth knowledge of hospital environments, Philips can provide a mix from its suite of services and software to enable hospitals to make improvements based on the information they are gathering from location data and location history.

The University Medical Center (UMC) in Tucson, Arizona, has already installed the Philips-AeroScout asset-tracking solution. The deployment covers the entire hospital comprised of eight floors and covering a million square feet and involves 2300 tagged assets, such as infusion pumps, beds, monitors, and wheelchairs.

## Case 8.13: PinnacleHealth Extends Asset Tracking to Community Hospital*

Pleased with the return on investment it is getting from the real-time asset tracking it deployed at its flagship 546-bed Harrisburg Hospital in Pennsylvania, PinnacleHealth is expanding the system to Community Hospital, a 148-bed facility it operates just outside of Harrisburg. When complete, the RFID-based system will let staff at both sites locate as many as 10,000 devices. In the future, biomedical engineering staff will be testing and installing active 433-MHz tags on at least 2000 assets and 200 readers at Community Hospital. The expansion adds to the patient- and asset-tracking RTLS from Radianse already installed at Harrisburg Hospital in the downtown area, used to track about 4500 assets, such as defibrillators, crash carts, and other critical life-support equipment. In addition to tracking assets, Pinnacle

*_Source_: Mary Catherine O'Connor. (2008). "N.C. Hospital Looks to RadarFind to Improve Asset Visibility." _RFID Journal_. Retrieved from http://www.rfidjournal.com/article/articleview/3878/1/1/.

Health has been using the system for more than three years to track about 23,000 patients annually in the surgical units at both Harrisburg and Community hospitals.

The organization also operates three other facilities: Polyclinic, an outpatient clinic with specialty hospital services in downtown Harrisburg; Seidle, a hospital in Mechanicsburg; and Cumberland, a physician office building and outpatient clinic, also in Mechanicsburg. Radianse's active 433-MHz tags and readers (which Radianse calls receivers) communicate via a proprietary air interface protocol. The readers are small box-shaped devices that connect to the hospital's local area network and relay the RFID data to a Radianse server. The device can receive a tag's signal from up to 50 or 60 ft away, and can pinpoint the tag's location within 3 ft. Radianse's software determines the tag's location based on the strength of the signal picked up by three or more readers.

Radianse had earlier upgraded the PinnacleHealth system by integrating Wi-Fi networking directly into the RFID readers so data can be transmitted via a wireless LAN. Previously, the devices had no built-in Wi-Fi network card and transmitted tag data via a cable connection either to a wired local area network or a separate Wi-Fi network bridge that then communicated wirelessly with a Wi-Fi access point. The Wi-Fi capability is something PinnacleHealth plans to use.

It is estimated that at least 70 percent of the readers installed will be Wi-Fi-enabled, if they perform well during testing. The two hospitals will share the server, so both facilities can see where assets are at either site. This is an important featurebecause the two hospitals share some high-cost specialty equipment, such as devices used in the lab for gastrointestinal endoscopy, hardware gets sometimes misplaced.

The RTLS will help care providers at both hospitals find all kinds of equipment more easily. When they need to find a specific item, they can access the Radianse software program from any computer at either hospital and conduct a search for a specific type of device, or even enter a device's serial number in order to locate a specific piece of equipment. Because the system tracks items in real time, it prevents the tendency by some to hoard items. The team also uses the system to find devices in order to inspect them or upgrade their software. By next year, PinnacleHealth hopes to be using a module from Radianse that will automate the manual-based inspection process and associated paperwork the hospital group currently uses. Instead of writing down which items have been inspected, the team will wear active tags equipped with buttons they can push to document where they are at what time as they finish inspections on equipment in a given area or room. The RFID data accumulated during the inspections can then be downloaded into a database used to track equipment inspections.

# CHAPTER 9

# Work-in-Progress Tracking

In addition to its use in tracking and locating assets, RFID is being used to track work-in-progress inventory.

## Case 9.1: Pro-X Pharmaceuticals Seeks RFID for Internal Benefits*

The nutritional supplement producer Pro-X Pharmaceuticals is deploying RFID to help track inventory and manage production. By deploying RFID technology, the company hopes to increase visibility into its manufacturing and inventory processes and thus able to respond better to fluctuations in demand (Fig. 9.1). Pro-X is an independent contract manufacturer that produces Roex-brand nutritional supplements and also provides manufacturing and packaging services for other companies marketing nutritional supplements.

Demand for nutritional supplements can shift quickly. For example, news reports related to bird flu caused spikes in demand for natural immunity boosters. On the other hand, reports that raise questions about the safety or effectiveness of any single supplement or ingredient used in supplements could quickly reduce demand. So Pro-X needs to be able to scale production schedules up and down quickly to maintain optimal inventory counts.

The products the company makes have natural ingredients and are aimed at preventing and decreasing health problems primarily associated with aging and disease. Despite its name, Pro-X Pharmaceuticals does not manufacture pharmaceuticals and because its products are nutritional supplements, they are not subject to U.S. Food and Drug Administration (FDA) approval or supervision.

---

**Source*: Mary Catherine O'Connor. (2006). "Pro-X Seeks RFID for Internal Benefits." *RFID Journal*. Retrieved from http://www.rfidjournal.com/article/articleview/2188/1/1/.

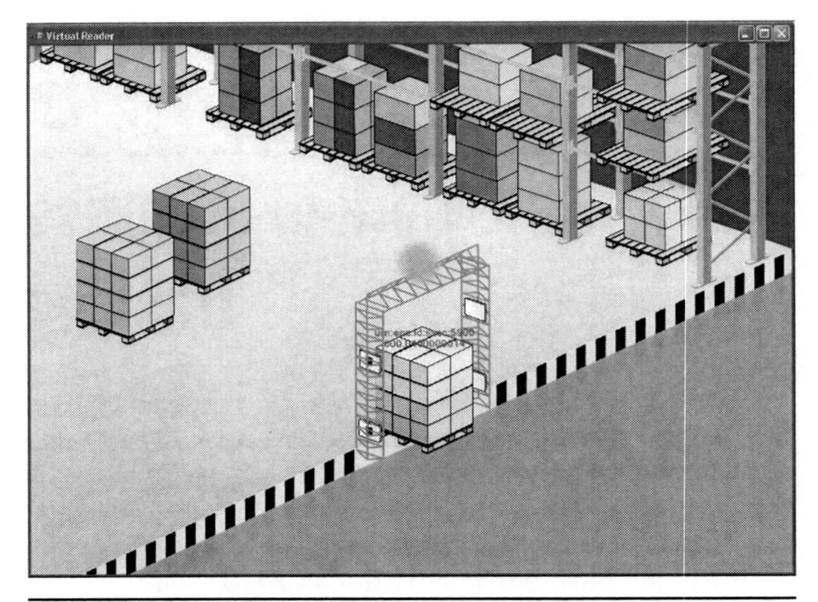


**Figure 9.1** RFID used to track and trace inventory and manage production of products. (*Source*: http://www.bridge-project.eu/data/Image/Portable%20 Demo.jpg)

Pro-X, based in Irvine, California, has chosen RFID systems integrator ODIN Technologies to help the company design and deploy RFID for work-in-progress manufacturing control and inventory tracking. The company is using passive, ultra-high frequency EPC Gen 2 tags and readers and middleware provided by Shipcom Wireless.

Pro-X had started investigating RFID as a possible means of improving product tracking, work-in-progress manufacturing processes, and product quality. It should be noted that the RFID deployment is not motivated in any way by retailer mandates or external pressure from Pro-X's distribution network. Roex sells products directly to customers through a mail-order system and also makes products for 232 retailers in the United States, none of which are currently using RFID for product tracking.

The designing phase of its RFID system is complete. Pro-X will add Gen 2 smart labels to bulk containers of ingredients used to make its products. At the Pro-X manufacturing facility, workers will apply smart labels to the containers of raw materials, which generally arrive in corrugated cardboard drums ranging from 5 to 45 gal in volume. The tags will be encoded, using a handheld interrogator, with unique IDs associated with a lot number, and any expiration dates linked to the ingredients.

Pro-X also performs quality tests on the ingredients received, to certify that they are safe and pure. The results of these tests will also be associated with the IDs. Before production of a particular item begins, the necessary tagged containers will be gathered and interrogated, and the tag data will be aggregated in a database. This production record will be associated with the bulk quantities of finished products, which will be placed in large tagged barrels. Encoded to each barrel's tag will be an ID correlating with the batch data, so the finished product can be linked to the raw materials.

Once each batch of finished supplement product is packaged into individual bottles for retail sale, those bottles will be packed in cases to which RFID smart labels will also be affixed. The tags will let Pro-X use interrogators in a number of different form factors, including handheld, fixed, and forklift-mounted devices, to track each case as it enters the inventory storage area and as it is picked to fulfill orders.

Pro-X believes that in addition to improving work-in-progress tracking and inventory management, RFID will help manage any product recalls they might have in the future. The company plans to install an RFID infrastructure in distribution facilities, for instance, enabling it to use the smart labels to receive the tagged cases of its products into inventory, to fulfill orders, and to automate the identification and collection of products that might be involved in a recall. Using RFID should lead to benefits for Pro-X's retailer customers because Pro-X will have more visibility into exactly where in the production process customers' orders are at any one time.

Pro-X has also recently installed Great Plains, a Microsoft platform for enterprise resource planning. The company is using the platform to manage accounting, inventory, and manufacturing processes. By pulling the tag data associated with the raw materials, finished products, and orders into the Great Plains applications, Shipcom will integrate the RFID system with Pro-X's back-end systems.

## Case 9.2: RFID Helps Endwave Track Work in Progress*

Endwave Defense Systems, a manufacturer of amplifiers, transceivers, and other RF communications modules for the aerospace and defense industries, is employing RFID technology to gain visibility on its production floor, and to monitor the status of work in progress at its 20,000-square-ft facility in Diamond Springs, California. By tracking each bin of components from the time a module is ordered through the final inspection of that product before its shipment, Endwave can know where a specific order is located, how long it has spent in any particular place, and who has worked on it. The company can also receive alerts if an unauthorized employee works on a product,

*Source: Claire Swedberg. (2008). "RFID Helps Endwave Track Work-in-Progress." *RFID Journal.* Retrieved from http://www.rfidjournal.com/article/articleview/4168/1/1/.

if a product is sent back to a previous workstation (or "cell") to be reworked, or if there is a delay at a certain point during production.

The firm piloted the system at the end of 2007, deploying it in the production facility in February 2008. When Endwave receives an order, production or shop floor employees take a plastic bin (bins vary in size from about half that of a shoe box to the equivalent of three shoe boxes) and fill it with all of the components necessary for the product's assembly. The bin passes through up to five separate assembly cells, then waits on a shelf until a worker takes it and begins working on the product. The employee then returns the pieces to the bin and passes it on to the next cell. Once the product completes assembly, it goes through quality assurance (including final visual inspection and physical tests) before being shipped to the customer.

With hundreds of products being manufactured at any given time in the large facility, if a customer calls for an update on a particular order, an employee must walk through the assembly floor to physically search for the corresponding bin. The RFID-based tracking system, designed by Omnitrol Networks, makes that process easier. Omnitrol installed 12 Motorola XR400 readers on Endwave's shelves, and at other locations where the bins are stored as they move through production. Endwave applied Alien Technology Squiggle EPC Gen 2 UHF RFID tags to the bins that pass through production, as well as on each employee's ID badge. Omnitrol software then instructs a Zebra Technologies RFID printer-encoder, located on Endwave's site, to create a badge by encoding its RFID tag and printing the unique ID number linked to the employee. When a customer calls in an order, a staff member enters an order number and the requested product type into the Omnitrol server, linking that data with a bin number. The employee then takes an empty bin, fills it with the necessary components for that particular product, and scans the bin's RFID tag, linking the order to that RFID number. As the bin passes to a production cell, it is placed on a shelf. The Motorola reader captures the bin's ID number and its location to within about 2 ft, then transmits the ID number—along with the time and date—to an Omnitrol application network appliance, which collects data from all of the readers and makes it available in a dashboard format, either on a video monitor attached to the appliance or on Endwave's computers.

The appliance can also send an alert via e-mail to Endwave management if a bin sits for too long on a particular shelf, or if it is taken from a shelf without being returned. What's more, the interrogators capture the unique ID number of each employee badge. That ID number, linked to a specific employee, is stored by the Omnitrol appliance, enabling management to view the history of any order, not only when and for how long it remained at a particular location, but also who had it and how quickly a specific process was completed. If an employee unauthorized to work on a particular order takes a bin from the shelf, the readers transmit an e-mail alert. Similarly, if there is a problem with an order and it must be sent back to a previous cell for reworking, management can receive an update indicating the product's location.

Endwave expects to see cost reductions comparable to at least one full-time employee. Another benefit comes from increased customer satisfaction due to faster response times to requests for order status reports. Endwave currently prints daily reports regarding work done on the assembly floor, which provide such analytics as revenue associated with a specific order. Previously, the company created such reports only about once weekly. This complete solution, including software, RFID readers, printers, installation, and training, costs less than $100,000.

## Case 9.3: RFID Illuminates Work in Progress for Neonlite*

Hong Kong energy-saving lightbulb manufacturer Neonlite Electronic and Lighting (maker of Megaman products distributed in more than 80 countries worldwide) is employing RFID and its own enterprise resource planning (ERP) system to manage product manufacturing, inventory, and shipment at one of its four manufacturing plants. The system, installed in January 2009 at its plant in Xiamen, China, utilizes RFID hardware from Intermec Technologies to track work in progress as trays of parts used to make the lamps move through the assembly line. GlobeRanger provided software for the deployment.

Neonlite is one of the fastest-growing firms in the lighting market. As business has expanded, so too has the manufacturing facility. To manage all of this growth, Neonlite decided to create a single, integrated management system through which it could gain visibility into inbound logistics, parts management, and work in progress. While most recent RFID applications have been focused on managing the supply chain for finished goods, Neonlite uses RFID within the entire production environment to improve the production material traceability and real-time inventory. Neonlite had already been utilizing Infor's ERP SyteLine warehouse management software to manage inbound parts and their consumption on the manufacturing floor. The new RFID system provides the company with greater visibility of the existing system, and extends that visibility through the manufacturing process by tagging trays used to carry and assemble parts and finished products and by tracking the trays' movement around the plant and warehouse.

---

*Source: Claire Swedberg. (2009). "RFID Illuminates Work-in-Progress for Neonlite." *RFID Journal*. Retrieved from http://www.rfidjournal.com/article/view/4691.

Prior to the RFID system's deployment, employees filled out transaction slips when parts were received or products were shipped, and then entered that data into the ERP system. With the RFID system, parts are loaded onto trays fitted with Intermec passive EPC Gen 2 RFID tags operating at 900 MHz. To indicate that a tray has been filled and is ready to be taken to an assembly station, a worker uses an Intermec handheld RFID interrogator to read the tray's tags. Each tag is then read once more as the tray passes an RFID portal (containing an Intermec IF5 fixed reader) on its way to the manufacturing floor, or into the warehouse for storage. The portal includes a motion detector that prompts the interrogator when a tagged tray approaches, at which time the reader begins interrogating the tag. As the tagged tray of parts passes from one work-in-progress phase to the next (often at a separate workstation), the tag is read via a handheld interrogator to update its status. Management can use the Infor SCM Warehouse Management software, integrated into the firm's ERP system, to view the tray's exact location and determine how long it has been there. Once assembly is complete, the finished lightbulbs are packed in tagged cartons and loaded on tagged pallets. The items then pass through the RFID portals several more times before shipment.

Infor SCM Warehouse Management and Infor SCM RFID play a pivotal role in Neonlite's inbound logistics operations. Infor SCM Warehouse Management enables the company to manage parts as they are moved into the warehouse or sent to another location, as well as when they are shipped from suppliers. Infor SCM RFID interprets the readers' data to enable Neonlite to locate and manage parts and product inventory on the manufacturing floor. The system improves production capacity by 15 to 20 percent, by ensuring the right material is available at the correct location when needed. With the SCM software, managers can determine how long it took for the parts to be consumed, and when additional items need to be ordered, as well as how much time the work-in-progress stage required, how long the finished lamps remained in storage before being shipped, and at what time they were shipped.

GlobeRanger's iMotion software platform was used to develop and test the RFID portals' processes, including the integration of motion sensors with the Intermec readers. The GlobeRanger software sends alerts (exception reports) that notify management via e-mail or by mobile phone if trays filled with work in progress are moved past the incorrect portal or at the wrong time, if too many or too few tray tags have passed a portal at a specific time, or if a tray's tag is not read at all. Additionally, the iMotion software will generate alerts if cartons are missing tags or by matching the ID number of each carton tag with that of the pallet's tag if the incorrect tagged cartons are loaded on a pallet. Installing the RFID solution took place in a number of stages.

First, a team with members from Neonlite, Infor, and GlobeRanger spent four weeks completing the conceptual design, during which it assessed Neonlite's needs and how the technology could address them. The group examined the physical containers (pallets and trays) and determined how they could best be tagged, and the group also studied the physical environment and flow of material and products on the manufacturing floor. The team decided that tagging work-in-progress trays that carry parts through the assembly process, as well as tagging cases and pallets, would allow the company to manage both production and logistics. For the next four weeks, the group developed and tested a prototype system. The companies installed a temporary portal, and users walked RFID tags through the process in order to test read rates. After that, the team began a two-week details design phase, drawing out the details of the installation, based on the prototype, including portal specifications, handhelds, and database design. The following four weeks were dedicated to developing, building, and testing the entire RFID system, then installing it and going live; a process that took an additional 12 weeks.

The system is still under partial deployment, to allow any minor adjustments before full-scale deployment. Neonlite also plans to install the system at its three other plants, though it has yet to determine a deployment date. In the meantime, the company hopes to see a positive return on the company's investment in approximately two years.

# CHAPTER 10

# Library Management System

RFID has already started replacing bar codes in the supply chain and is used for a wide variety of inventory tracking applications. A major potential growth area for RFID is expected to be in libraries due to decreasing library budgets and decreasing consumer purchasing power. It seems likely that more people will be borrowing books instead of purchasing them.

In a typical inventory management system, the RFID tag is used as a "throw-away" technology; as soon as the item reaches the customer, its life cycle ends. However, the use of RFID to identify a library book allows the tag to be used multiple times since the same book remains in circulation for a long time (i.e., it is checked in and out a number of times).

In general, a library RFID management system will consist of four phases (see Fig. 10.1). The first phase handles the library security system only, with the RFID tag (Fig. 10.2) replacing the electromagnetic (EM) security strip for detecting books leaving the library without permission. The RFID tag can be used as a security mechanism by incorporating a security bit with the values of "checked in" and "checked out." Alarm systems placed at the exit gates ensure that only checked-out books leave the library (Fig. 10.3).

The second phase supports the library circulation process by reading the book's access number from the RFID (and not a bar code) (Fig. 10.4). The primary advantage is that reading the RFID tag does not require line of sight and allows for multiple items to be checked in or checked out simultaneously.

The third phase is a self check-in and checkout process. This will allow for faster patron check-in and checkout that frees up the staff to provide better customer service. Indoor and/or outdoor book-drop stations (Fig. 10.5) can be used for the check-in process.

LibBest library RFID management system

Shelf management
Check-in/out service
Self check-in/out
Book drop
Tagging
Antitheft detection

**Figure 10.1** Best practice for a library management system. (*Source*: http://www.rfid-library.com/en/default_e.html)

**Figure 10.2** Most commonly used RFID tag for books. (*Source*: © Bibliotheca RFID Library Systems)

Finally, inventory of the books can be conducted quickly in the fourth phase. An added benefit is the ability to locate missing books and books that have been misplaced on the shelf.

The following criteria are relevant for the choice and comparison of RFID systems in libraries:

- The functions are performed with high reliability (reading in a sensor gate, self-check, staff station, inventory control, return station).
- The availability of compatible chips (RFID labels) must be guaranteed.

**Figure 10.3** Alarm systems can be placed at the exit gates to ensure that only checked-out books leave the library. (*Source*: http://www.rfid-library.com/en/default_e.html)

**Figure 10.4** Placement of an RFID tag. (*Source*: http://www.rfid-library.com/en/default_e.html)

- Chip technology is compatible with different generations of RFID systems from different producers (ISO 15693 Standard).
- Additional security strips are not necessary (no electromagnetic strips).
- The system can be extended with ID cards for access control to a lobby room, payment at a copy machine, Internet access, coffee machines, and other services.

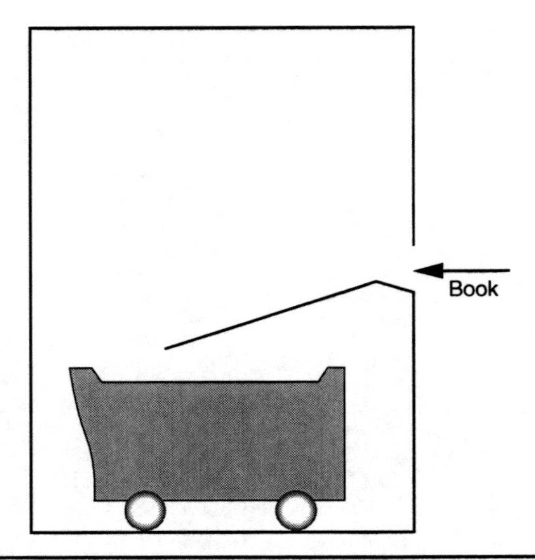


**Figure 10.5** Indoor or outdoor book drop. (*Source*: http://www.rfid-library.com/en/default_e.html)

- Support must be given by the installed management system software (circulation software); SIP2 is the standard today, and NCIP is upcoming.
- The system allows for all audio and audiovisual media to be equipped with RFID labels (only exception: double-sided DVDs).
- Additional servers are not required and the system components can be exchanged during operation of the library.

Libraries are institutions that do not measure "profits" as their success. They are expected to budget the return on investment (ROI) toward creating new services or improving existing services. They are also expected to allocate their ROI to offset possible budget cuts or loss of buying power when budgets do not keep up with inflation. It is demonstrated that the greatest return on RFID investment will be achieved by practically eliminating the need for the circulation staff. It is essential that ROI takes into account the measurement of customer satisfaction. Self checkout systems can be viewed as a burden by customers who feel that libraries have eliminated the "human factor" and have shifted the burden of checkout to the users themselves. Or it can be viewed as an "ATM" to provide faster service.

Although RFID could be very promising to identify books and media (including the packaging of audiovisual media for identifying its contents), there may be problems in implementing RFID for items such as any magazines, pamphlets, and music sheets that do not have a good location for a 2-inch square tag. Also, the tag costs for these would be significant, especially for magazines that have a short circulation life. To have an alternate approach to maintain the inventory for such items will increase costs significantly, which may act as a deterrent to embracing RFID.

Libraries also need tags that are durable because the same tag would be used repeatedly. The library function may need the longest-lived tags because books can be circulated for decades. Approximately 70 million library books were RFID-tagged worldwide by 2004, with hundreds of libraries all around the globe using RFID systems.

## Case 10.1: National Library Board Singapore*

The biggest RFID installation in the world is by the National Library Board (NLB) in Singapore. In 1994, a "Library 2000" vision document was presented to the Singapore government identifying the need for an effective library management system to position Singapore as an information society. The library should make available knowledge from across the world to help patrons benefit by bridging knowledge gaps, termed "global knowledge arbitrage," the document argues. The NLB provides services for over 70 libraries in Singapore, which includes 39 national libraries, 18 community libraries, and 18 children's libraries. It has two logistics operations in place: Library Supply Services that handles the supply services and the Networks Operations Center that directly serves the customers.

With a mission toward expanding the learning capacity of the nation, NLB set about the task of increasing annual book loans from about 10 million in 1994.

### Need for RFID System

The bar code system in use at the libraries at that time was difficult for library users. To borrow a book, library users had to carefully align the book with the bar code reader before the machine could read the bar code at all. Meanwhile, book returns were handled manually. To speed up returns, the NLB had introduced book return chutes. These return chutes were located at library entrances and allowed users to drop off books any time of the day. However, library users still had to wait for the staff to update the loan records in the system. For example, library users could return books over the weekend when the library was closed, but until the library staff updated the overnight pileup of books in the chute against the loan records of

*_Source_: Devadoss, P.R. (2010). "RFID and Organizational Transformation in the National Library Board of Singapore." _World Scientific Books_, pp 1–26.

the respective library users, the library users could not check out other books.

Exploring for a better technology to handle book loans, the NLB identified RFID as a potential solution. In Singapore, ST Logistics had been exploring the use of RFID for logistics operations for a couple of years. Furthermore, its technology partner, ST Electronics, held the RFID expertise in Singapore. The NLB saw a similarity between its operations and the logistics business. Since the operational aspects were similar, the CEO of ST Logistics (which has since become Sembawang Logistics) invited the head of ST Electronics to discuss the potential of RFID for library use. Along with the NLB, the partners worked together to develop a prototype for library use. A demonstration was conducted in November 1997.

## Deploying RFID

The case use of RFID tags on all books at the NLB was a key project that had the NLB's assistant CEO as project sponsor. A number of other project teams carried out other service developments, layered over the RFID project. To tag a book, an RFID chip was embedded in the spine of the book (currently, with a much smaller chip available, it is pasted on the last page of the book), allowing scanners to identify the book in close proximity. The chip used the signal from the scanner to power a response returning the data embedded in the chip, a design known as passive RFID technology. All NLB library items are now tagged with an RFID chip containing information pertaining to the book, the library branch to which the book belongs, and the number of the rack where the book is shelved. RFID scanners read the data stored in an RFID chip to identify the library item. In a book loan or return process, the data is used together with the library user's identification to manage the library user's loan information. The data is initially stored in a local server, which operates with a backup and is then synchronized with the centralized data servers.

The RFID technology made checking out books easier. The checkout counters, called borrowing stations, were designed with a simple interface offering options for the four official languages of Singapore (English, Chinese, Malay, and Tamil). Users could log into the system by placing their identity cards into the machine. The users could then proceed to place each book they wished to check out on the reader, and the screen would confirm the loan by displaying the title of the book being checked out and the loan record status of the library user.

The RFID-enabled book-drop chute is now a feature at every NLB library. Because it is located at the entrance of the library, library users can return books any time of the day. An advantage of the RFID system is that it allows instantaneous update of the user's account, enabling the immediate renewal of the user's loan quota. This instantaneous update is achieved by placing an RFID scanner in the book-drop chute.

At the book drop, the user drops the book in the chute, and the RFID scanner updates the user's book loan records in the system instantaneously. The efficiency of checking out books and returning them at book drops at any library has improved user experience at libraries, further helping in the growth of book loans at NLB.

The whole implementation was again piloted at the next library due to reopen after renovation, namely the Toa Payoh Community Library, in 1999. Revised versions of the system had been piloted at two more libraries before the system was functioning to the satisfaction of the NLB. The success of the technology during pilot testing prompted other libraries to request the system. And in 2000, the NLB invited global tenders (bids) to implement the system across all its libraries. ST Logitrack was awarded the tender and has since rolled out the RFID systems in all NLB libraries in Singapore. The entire process was completed in April 2002.

The NLB's adoption of RFID was essentially an instance of information technology (IT) deployment to achieve organizational goals. RFID demonstrated potential in removing queues, delivering better service quality, and giving employees more time for value-added tasks. Additional, IT made innovation of new services possible at libraries. With the improvements, growing loans also meant an increasing number of returns, thus placing a heavy burden on the staff handling the shelving of books. The NLB now employs part-time workers who help the regular staff with shelving. This strategy helps the NLB carve the tedious work process into smaller, manageable schedules allocated to the part-time staff. In addition, several community programs have been implemented to bring in volunteers to help shelve books. Such programs also benefit the NLB by helping it reach out to the community and engage them in its daily work process.

A negative impact of the adoption of IT at the NLB was job insecurity, which was a growing concern at the NLB when the new system was introduced. It became evident that the new systems provided immense savings in terms of manpower in the organization, and the staff was concerned that it would mean the loss of jobs for some. This fear was felt particularly among those who had little knowledge of information technologies. The changes at the NLB were seen as a shift in the culture of the people within the organization. Such a shift was also viewed as necessary to the NLB in its growth and ability to deliver excellence in its services. The management positioned the shift in the organization as a value proposition to help increase productivity and redefine routine job tasks, but they recognized that some might query the change. Top management engaged the staff in dialogue and conveyed the message that the newly introduced technologies were meant to help increase productivity. The NLB also provided training sessions, opportunities for skill development, and redeployment of some staff to other jobs. The CEO made frequent visits to all libraries and met with the staff and held tea sessions where staff aired their suggestions and concerns.

## Transforming the Organization

In addition to the adoption of IT, the NLB also adopted a lifestyle approach in designing the library environment, thereby changing the perception of a library. This approach meant locating libraries in shopping malls to make them accessible to users, setting up a cafe within the library, and changing the ambience of a library from the traditional somber one to a more vibrant atmosphere to attract visitors. The lifestyle concept changed the nature of libraries in Singapore.

Today, NLB libraries are cozy places where visitors can browse a variety of book and multimedia collections and tap into various services amid plush surroundings. The libraries are also equipped with Internet terminals and multimedia kiosks. Digital resources are available through terminals at the library, as well as the e-library hub (www.elibraryhub.com), which complements the NLB's existing services. At the NLB libraries, users with their own laptops and personal digital assistants (PDAs) can tap into broadband Internet service from a private vendor, which includes access to the NLB's digital libraries.

With increasing adoption of RFID technology at more branches, book loans at the NLB and library user visits to the various NLB branches grew annually. The increased productivity was managed with retrained staff from other functions that had become redundant due to the introduction of IT. As RFID was adopted at each new library with more services that were automated, fewer employees were needed to staff a library. The NLB countered this by increasing the responsibilities of lower rank staff to the extent that the first fully self-service library was launched with just one Systems Library Officer and one concierge. This minimally staffed library manages approximately 2000 loans a day. At the end of 2002, NLB's annual loans were over 32 million, and its collection numbered approximately 8 million, including books and multimedia material. It has had about 31.7 million visitors and 2.1 million memberships, and it handled 1.8 million inquiries in the year. The NLB estimates that given its over 30 million loans per year and less than a minute per transaction at the counter service at present, it would need to add 2000 more staff to its workforce to keep up current service levels without the RFID system.

Through the adoption of technology in its various services, the NLB is now equipped to quickly deploy loan services even at remote community events, thus taking the library to the people. This service works by connecting to the library network using a laptop and a virtual private network. The computer is attached to a scanner, which reads the RFID and logs the loan. RFID-tagging its collections has also helped NLB drastically reduce the time spent in stocktaking. None of its libraries now close for stocktaking, and the entire exercise at a library is completed overnight, except for the anomalies in reports which are followed up later. NLB is pursuing a change in RFID chip technology to further improve the efficiency of the system.

## Case 10.2: Belgian University Library*

In Leuven, the biggest Belgian University library and the first university library in all of Europe (Katholieke Universiteit Leuven, Campus Library Arenberg) uses a Bibliotheca RFID system to increase the efficiency and quantity of media circulation. The IT engineers of the KU Leuven were the first to program library management software (LMS) for university libraries, as a program called Dobis/Libis. The Bibliotheca RFID system was chosen based on input from the staff as well as the layout and the architecture of the library. The library is open 14 hours a day throughout the week, and the total book collection is 4.5 million in 18 departments and two facilities. At present, there are 100,000 audiovisual media items housed in a new library building. In the future, this library will contain about 250,000 media items for public access; an additional 600,000 are located in compact shelves. Use of an automatic book-sorting machine is planned. In order for this library to function with a limited staff of 20 people in two shifts, new technology for self-service checkout and check-in of books is necessary. This means that only three to four staff people are present in the library at any one time. They are responsible for the management and adding of new items and assisting patrons.

## Case 10.3: Winterthur Libraries†

In Winterthur, Switzerland, three libraries are working with RFID. The most recently opened public library holds a collection of about 250,000 items. The main reason to implement the RFID system was that the number of borrowed items was steadily increasing, especially with new media like CDs and DVDs. In order to keep the number of staff at the same level and at the same time offer a much wider and more attractive collection, the decision was clear to use RFID in order to automate all possible work processes. With the implementation of the RFID system, the tasks of the librarians have changed.

---

**Source*: Bibliotheca RFID Library Systems AG. (2010). *Radio-Frequency-Identification for Security and Media Circulation in Libraries.* Hinterbergstr: Bibliotheca RFID Library Systems AG.

†*Source*: Bibliotheca RFID Library Systems AG. (2010). *Radio-Frequency-Identification for Security and Media Circulation in Libraries.* Hinterbergstr: Bibliotheca RFID Library Systems AG.

More and more younger people use the facilities and so the staff has to take on a survey and social function—which leads to the necessity of a self-issue station, in order to give staff time to look after the young visitors. The library lobby contains four book return stations and can be accessed 24 hours per day with an RFID ID card. In the next (inner) area, four self-issue stations are installed, plus four staff stations. All books and CDs are tagged with RFID. The RFID ID card also contains a chip with payment function, to be used at the coffee and copy machine. There are various plans for extended uses of the cards (e.g., Internet access). The entrance to the inner area is secured with RFID gates.

## Case 10.4: Vienna Public Library*

The Vienna public library, with an inventory of 300,000 items, has opened in an entirely new building. There are 240,000 books being tagged with RFID labels, plus 60,000 CDs and DVDs being tagged with special CD-RFID labels. The RFID system is of vital importance to allow the library to be run by a limited number of well-educated staff who will have time to attend to visitors. The system comprises 13 staff stations, 11 gate antennas, and 5 self-check stations. Both the staff stations and self-check stations are able to read tags on a stack of books simultaneously. The system is used by 3500 visitors daily.

**Source*: Bibliotheca RFID Library Systems AG. (2010). *Radio-Frequency-Identification for Security and Media Circulation in Libraries*. Hinterbergstr: Bibliotheca RFID Library Systems AG.

# CHAPTER 11

# Returnable Asset Tracking

Reusable assets are often misplaced, and the lack of visibility into their movement can lead to substantial losses for companies. RFID asset-tracking solutions enable companies to better manage their returnable assets. It can provide information about assets due back from various trading partners, and also presents information regarding the status of returnable assets against associated order numbers, improving the visibility of assets possessed by different partners across the supply chain.

## Case 11.1: Australian Companies Say Pallet-Tracking Project Proves RFID's Mettle*

In July 2007, an Australian consortium of businesses and organizations completed a two-month pilot-test of EPC Gen 2 RFID tags and interrogators affixed to wooden pallets. The results proved RFID can raise productivity and efficiencies in a multi-industry supply chain. The pilot, known as the National EPC Network Demonstrator Project Extension, was managed by GS1 Australia, a branch of the international standards-setting organization GS1, in cooperation with RMIT University in Melbourne, Australia. The purpose of the pilot was to demonstrate that the EPC/RFID is not merely a theory that works on paper. Current supply chain practices involve quite a bit of paperwork, often by the sender, the receiver, and the transporter. Companies now want to remove counting, data entry, and paperwork, and still ensure 100 percent accuracy and integration with their enterprise resource planning (ERP) system.

For the pilot, CHEP, a global pooling service for containers and pallets, provided wooden pallets fitted with passive EPC UHF RFID tags

---

**Source*: Beth Bacheldor. (2007). "Australian Companies Say Pallet-Tracking Project Proves RFID's Mettle." *RFID Journal*. Retrieved from http://www.rfidjournal.com/article/view/3467

**Figure 11.1** CHEP pallets fitted with RFID tags. (*Source*: http://mams.rmit.edu.au/2lv9x2qhyxom1.jpg)

(Fig. 11.1). Other participants included office products supplier ACCO Australia, consumer packaged goods (CPG) provider Capilano Honey, logistics provider Westgate, discount supermarket operator Franklins Australia, CPG company Procter & Gamble, supply chain and logistics provider Linfox, and CPG company MasterFoods. Service providers Telstra and Retriever Communications also contributed to the project.

The 3300 tags employed in the pilot were passive EPC Gen 2 tags from Impinj. The standard used for the unique numbering of the pallets was a Global Returnable Asset Indicator (GRAI), a GS1 numbering structure for returnable assets. A Global Location Number (GLN), another GS1 numbering structure, was utilized to uniquely identify locations where handheld readers were operated. Telstra's EPCIS-compliant Adaptive Asset Manager (AAM) software was used to manage and share the RFID data collected during the pilot, which participants were able to access via a Web interface. The data was also communicated to handheld personal digital assistants (PDAs) used by CHEP truck drivers. The frequency employed in the pilot ranged from 920 to 926 MHz (UHF), the Australian UHF RFID band, using a combination of 1 and 4 watts of radiated power (fixed readers transmitted signals of up to 4 watts of power, while the handheld scanners operated at 1 watt).

After accessing an order via their PDA, CHEP truck drivers read the tags while picking up the ordered pallets (Fig. 11.2) and loading them onto trucks for delivery to Capilano Honey, P&G, ACCO, or MasterFoods. As each pallet passed a fixed RFID reader, its tag information was sent to Telstra's AAM software, then relayed (via

**Figure 11.2** CHEP truck drivers read tags while picking up multiple pallets for a customer order. (*Source*: http://mams.rmit.edu.au/2lv9x2qhyxom1.jpg)

GPRS-protocol mobile phone service) to the driver's PDA. At that point, an indicator light on the PDA software changed to yellow status, showing that pallets were being read for that order. Once all the pallets were read, matching the quantity expected for the order, the PDA indicator turned green, after which the driver departed to deliver the pallets to the customer.

Fixed RFID interrogators deployed at the customers' sites also read the tagged pallets as they were being unloaded from a truck, enabling the system to confirm individual pallets in a given block of pallets. Once again, Telstra's AAM system passed the information to the driver's PDA, which indicated a yellow status light to indicate the pallets were being read. Once all pallet tags were read, the PDA indicator turned green. If there were no discrepancies, the driver could close the order. If a discrepancy existed, it could be resolved there and then, rather than waiting for a subsequent reconciliation process.

A returns process was also tested. Westgate retrieved empty pallets from Franklins and read their tags as they passed through its facility and were taken to CHEP's warehouse. There, the pallet tags were read once more.

The real key to success, though, was the way this data was integrated into the system. Orders were entered into the system and deliveries scheduled for trucks, and this was the last manual data entry in the process. With accurate reads occurring 100 percent of the time, the pilot was able to demonstrate successful electronic proof of deliveries (ePODs). Each pallet has a unique number, so every pallet can be accounted for individually. This ensures that reconciliation is done to the point of full electronic proof of delivery. Instead of having to write anything down and rekey it, everything is done electronically.

Proving that RFID can be used to get a 100 percent read rate allows RFID data to serve as proof of delivery and takes away any uncertainty. The use of RFID also removes the manual checking processes, so there are additional benefits in process efficiency. The pilot participants were able to capture and share information at different points in the supply chain. It gave them visibility of goods through the supply chain and turned indiscriminate EPC RFID reads into business transactions. Some of the customers in the pilot reported productivity gains of 14.3 and 22.2 percent, achieved by reducing process times and by using ePODs rather than paper-based processes. CHEP estimated productivity gains of 28 percent for the entire end-to-end delivery process.

The most difficult challenge was achieving 100 percent reliability. Read rates were hampered by the condition of the pallets' wood, as well as their paint. The pilot used both new and used pallets, and some of the new pallets had been painted the day before the trial. The moisture content in the new pallets was much higher than that in the conditioned ones, which had dried with age. The discrepancies had to be addressed when tuning the tags and insulating the tags from the wooden substrate. To resolve this issue, a couple of millimeters of foam were adhered to all pallets that were used, along with foil on the newer pallets. Once these obstacles were overcome, however, the pilot's objectives were met. RFID really did deliver on its promise, even though it was only for the last few runs.

## Case 11.2: Rewe Deploying RFID Long-Range, Real-Time Locating System*

Rewe Group, a European retail and food group that reported approximately € 50 billion ($72 billion) in revenue in 2008, completed a test in which it employed a radio frequency identification real-time locating system (RTLS) to track returnable transport items (RTIs) at dock doors at its distribution center in Buttenheim, in southern Germany. The company now plans to focus on integrating the RFID application into its warehouse-management system so that it can begin a pilot program leading to its eventual expansion of the application. Rewe, the third largest food retailer in Europe and the second largest in Germany, operates 29 distribution centers (DCs) and 10,305 stores in that country, as well as another 4409 retail outlets in 14 other European nations. The company's German supply chain operations use about 36 million pallets and roll containers annually. Rewe had previously tested RFID-tagged pallets at its distribution center in Norderstedt, in northern Germany. The Norderstedt application involved RFID receiving portals, forklift interrogators, and handheld

*Source: Rhea Wessel. (2009). "Rewe Deploying Long-Range Real-Time Location RFID System." *RFID Journal*. Retrieved from http://www.rfidjournal.com/article/view/5187

readers for interrogating EPC Gen 2 passive ultra-high frequency (UHF) tags on pallets.

In Buttenheim, Rewe installed a system that can read tagged RTIs at 105 dock doors in a distribution center that handles dry foods and perishable items. Once the system is integrated, the company plans to use it to improve management of RTIs. By individually identifying the items instead of tracking them in bulk as the company does at present, Rewe can avoid the loss of RTIs and improve the process of accounting for the items, thus saving money. Rewe considered conventional EPC Gen 2 RFID interrogators for the application, but ultimately decided on the Mojix STAR system, which has a real-time locating feature allowing for the identification of passive EPC Gen 2 tags over a long distance. The system can also determine the tags' location at different depths. This feature was particularly valuable in the shipping area, where Rewe needed to know which RTIs were loaded on trucks and which ones were still staged at the dock door. The Mojix system was able to distinguish between the two after the system was fine-tuned. Rewe is working to improve the visibility of its transport containers. Bar codes are an important part of the container-management process, but Rewe sees RFID as part of the solution for managing RTIs efficiently, given the ease and speed of identification with the technology. In addition, RFID would allow Rewe "process security" regarding RTI tracking, since it would eliminate errors inherent to bar code scanning. The test of Mojix's hardware began in February 2009, and involved several thousand tagged RTIs of different types, including pallets, roll containers, and frozen-food containers composed of metal and plastic. Some 16 dock doors were initially outfitted with the technology in the early stages of the test of RFID-enabled dock-door shipping processes.

The test was conducted with UPM Raflatac's DogBone tags inside polycarbonate casings that were bolted to the RTIs. Rewe is presently evaluating the tags and casings for their ability to withstand impact. Later, after Rewe has fully implemented and expanded the application, the company may ask its RTI suppliers to deliver the items tagged. Once the application is fully integrated into Rewe's warehouse-management system, Rewe will be able to automatically compare the RTIs on a particular truck with those that should be on that vehicle. Upon entering the distribution center, a driver will be directed to a specific dock door for loading. As the driver moves a tagged RTI from the staging area onto his truck, the RFID system will identify it and a visual signal (a red or green light) will let him know if he has moved the correct item onto the vehicle. Since an individual store is responsible for the RTIs and is liable for those not returned, Rewe could utilize the system to efficiently generate an overview illustrating which RTIs have been sent to which stores. Currently, employees at a store scan the bar code label on each RTI as it arrives or is sent back to the DC, thus booking it into and out of the asset-management system. In the future, when the RTIs are returned to the distribution center, their RFID tags will be interrogated, enabling Rewe to know the full extent of its RTI inventory on site and to confirm that items were indeed returned. Someday, if the system is expanded to individual stores, a store could employ RFID to identify the RTIs. But Rewe is initially focused on implementing the Buttenheim RFID pilot, which is expected to commence in October of the same year.

# PART IV

# Summary and Looking Ahead



This final section summarizes the book's topics and then takes a look ahead for RFID usage in supply chain management applications. While RFID technology is certainly not new, the increasing attractiveness of this technology is undeniable. RFID has been reshaping the supply chain management landscape as it moves toward an RFID-integrated supply chain. It is unreasonable to believe that RFID will totally replace the bar code; however, it will be a complementary technology to improve end-to-end supply chain management.

# CHAPTER 12

# Summing It Up and Looking Ahead

There are three levels of motivation for a company to choose to implement RFID. The first is simply a reactive implementation to comply with a trading partner's request (or mandate). Second, a company may elect to implement RFID to improve efficiencies for a specific process within the company. Third, the implementation could be to improve efficiencies of multiple connected processes within the company or it may involve using RFID across the entire supply chain. At any of the levels, a company will have to overcome challenges in order to achieve the desired benefits.

This chapter summarizes the challenges and benefits of using RFID in the supply chain, as described in the case studies in preceding chapters. Understanding the trade-off between the benefits and challenges leads to the desired organizational performance (typically measured as the return on investments). This chapter then concludes with a look at what is next for RFID in supply chain management.

## Challenges

The primary issues for designing RFID applications are operational, technical, and of course, financial challenges. In addition, security and privacy challenges are a growing concern.

Operationally, as companies are being required to label cases and pallets and sometimes items, a challenge is positioning tags, especially on individual items because those tags must be read within a case or pallet. Technical challenges begin with the lack of consensus on standards. While international standards already exist, establishing truly global standards is the real challenge. Financially, cost is a major factor in determining the speed at which RFID technology is adopted. An RFID system requires expenditures not only for tags, readers, hardware, and software but also for system maintenance and training. Like many new technologies, RFID systems may create security and privacy challenges. The organization's security may be at risk, and consumer privacy concerns have also surfaced.

## Benefits in the Supply Chain

Supply chain managers are constantly pressured to focus efforts and investments on initiatives aimed at reducing costs and improving customer service. With the current global economic problems, supply chain applications will remain a source of needed investment to improve performance. RFID technology has the potential to improve supply chain performance. The cases presented in the prior chapters represent a sample of the many benefits that have led companies to pilot-test and implement RFID systems.

Companies are realizing that implementing RFID is beneficial not only at the end points of the supply chain, such as point-of-sale customer contact points, tracking promotions, displays at the end of a retail store aisle (often referred to as end-cap displays), or electronic proof of delivery, but also for conducting root cause analysis. These companies are analyzing data from RFID systems, point-of-sale terminals, and other sources to determine not just out-of-stock situations but also other information to improve operations.

RFID provides persistent, real-time identification information with minimal human intervention, allowing more frequent data collection and greater information capture. With RFID, a dock door, conveyor, forklift, or workstation becomes an important data-collection instrument that can read and help reconcile the location and status of goods in the supply chain (Fig. 12.1). RFID offers unprecedented levels of data reliability and intelligence that can be used to eliminate

**Figure 12.1** Data collection on location of goods and updating inventory position status. (*Source*: http://d2eosjbgw49cu5.cloudfront.net/rfid-weblog.com/imgname-worlds_first_rfid_technology_for_measuring_the_distance_between_uhfband_antenna_and_ic_tags_courtesy_omron—50226711-Omron-RFID-Technology.gif)

waste, target resource reductions where the business will benefit most, align manufacturing with business priorities, manage trading partners, maintain high levels of customer service, gain supply chain agility, and more.

Companies that can successfully use this data to identify and fix problems within the supply chain will gain the greatest benefit and the most competitive advantage from RFID. Thus, RFID applications increase supply chain visibility by enabling a deeper understanding of the business processes, resulting in significant savings.

Real value is derived when RFID-delivered operational data is tightly integrated with enterprise resource planning (ERP) systems. This ensures that the company's business planners, operations managers, customer service staff, and sales and finance departments all have real-time visibility into what is actually taking place at remote locations. All relevant functional areas will have concurrent access to the same high-quality data.

As seen in the cases presented in the earlier chapters, the benefits include improved supply chain visibility, productivity, customer service, and asset management. Supply chain visibility leads to reduction in overstocking as well as reduction in out-of-stock occurrences. Improved productivity is largely due to process automations as well as the enhanced accuracy and availability of information. The following summarizes the benefits of RFID in five broad areas:

- *Automation.* Reducing manual processes through automated scanning and data entry improves productivity, thus allowing resources to be reallocated to higher-value activities.
- *Integrity.* Improving the integrity of real-time supply chain information, with increased authentication and security and tracking capabilities, reduces errors, shrinkage, and counterfeiting while improving customer satisfaction.
- *Velocity.* Minimizing the time spent on finding and tracking needed assets increases product flow and handling speeds.
- *Insight.* Providing real-time information makes possible faster, better-informed decisions and the ability to be more responsive to the customer.
- *Capability.* Providing quality enhancements, process improvements, and new applications helps to meet demands of supply chain partners and enhance customer experiences.

Understanding the benefits and the challenges is important for organizational performance. While a favorable return on investment is desired, sometimes direct financial measures are not the lone factor. Improved customer service also affects the bottom line and should be considered.

**FIGURE 12.2** Store of the future. (*Source*: http://www.pinoytechblog.com/wp-content/uploads/rfid_wagon.jpg)

## So, What's Next?

The store of the future has been depicted as one where a customer can walk the aisles, collecting goods into a shopping cart, and then walk out the door without human intervention verifying the purchases (Fig. 12.2). The consumer's purchases are automatically registered with the use of RFID, and a cashless payment is charged to the customer's primary credit (or debit) card on record.

This scenario assumes every consumer packaged good will be tagged with an RFID tag. Although the cost of implementing an RFID system is on the decline, using item-level RFID tags in the retail supply chain is not reasonable given today's RFID tag costs. However, it is reasonable to foresee RFID tagging in use at the case and pallet levels for all consumer packaged goods within the next 10 years.

Many challenges remain for item-level tagging for all consumer packaged goods. Besides the cost of the tags, the challenge of automating the application of tags to individual items must be met.

Beyond consumer packaged goods, RFID applications are more likely to be adopted in service operations, such as health care and logistics, for improving customer service and reducing overall operating costs. By providing the dual benefits of improved customer service and a lower operating cost for the company, RFID technology can give a company a competitive advantage.

As more companies adopt RFID for their organization, the cost will continue to drop. This in turn, will allow more companies to enter the RFID field, a cycle likely to continue for some time.

# Bibliography


Angeles, R. (2009), "Perceptions of the Importance of Absorptive Capacity Attributes as They Relate to Radio Frequency Identification Implementation by Firms Anticipating Radio Frequency Identification Use," *International Journal of Management and Enterprise Development*, 6(1):88–117.

Angeles, R. (2007), "Empirical Study on the Critical Success Factors for RFID Implementation and Their Relationships with Expected Deployment Outcomes," *Proceedings of the 38th Annual Meeting of the Decision Sciences Institute*, Phoenix, AZ, November 17–20, pp. 1631–1636.

Angeles, R. (2005), "RFID Technologies: Supply-Chain Applications and Implementation Issues," *Information Systems Management*, 22(1):51–65.

Bacheldor, B. (2009), "Jackson Memorial Enlists Thousands of RFID Tags to Track Assets," *RFID Journal*, Retrieved from http://www.rfidjournal.com/article/view/4638.

Bacheldor, B. (2009), "Philips Introduces Asset-Tracking System for Health Care," *RFID Journal*, Retrieved from http://www.rfidjournal.com/article/articleview/2869/1/1/.

Bacheldor, B. (2008), "AeroScout Unveils New Asset-Tracking Platform," *RFID Journal*, Retrieved from http://www.rfidjournal.com/article/articleview/3887/1/1/.

Bacheldor, B. (2008), "PinnacleHealth Extends Asset Tracking to Community Hospital," *RFID Journal*, Retrieved from http://www.rfidjournal.com/article/view/4351.

Bacheldor, B. (2007), "Australian Companies Say Pallet-Tracking Project Proves RFID's Mettle," *RFID Journal*, Retrieved from http://www.rfidjournal.com/article/view/3467

Bacheldor, B. (2007), "Carolinas HealthCare System Deploying RTLS at Its 20 Hospitals," *RFID Journal*, Retrieved from http://www.rfidjournal.com/article/articleview/3704/1/1.

Bacheldor, B. (2007), "Denver Health Adopting a Hospital-Wide RTLS System," *RFID Journal*, Retrieved from http://www.rfidjournal.com/article/articleview/3718/1/1/.

Bibliotheca RFID Library Systems AG. (2010), *Radio-Frequency-Identification for Security and Media Circulation in Libraries*, Hinterbergstr: Bibliotheca RFID Library Systems AG.

Bottani, E. and Rizzi, A. (2008), "Economical Assessment of the Impact of RFID Technology and EPC System on the Fast-Moving Consumer Goods Supply Chain," *International Journal of Production Economics*, 112(2):548–569.

Cannon, A., Reyes, P.M., Frazier, G.V., and Prater, E. (2008), "RFID in the Contemporary Supply Chain: Multiple Perspectives on Its Benefits and Risks," *International Journal of Operations and Production Management*, 28(5):433–454.

Cannon, A., Reyes, P.M., Frazier, G.V., and Prater, E. (2006), "How Much New Theory Do We Need at the Dawning of the RFID Era?" *17th Annual Conference of the Production and Operations Management Society*, Boston, Massachusetts, April 28–May 1.

Choy, K.L., So, S.C.K., Liu, J.J., Lau, H., and Kwok, S.K. (2007), "Improving Logistics Visibility in a Supply Chain: An Integrated Approach with Radio Frequency Technology," *International Journal of Integrated Supply Management*, 3(2):135–155.

de Kok, A.G., van Donselaar, K.H., and van Woensel, T. (2008), "A Break-Even Analysis of RFID Technology for Inventory Sensitive to Shrinkage," *International Journal of Production Economics*, 112(2):521–531.

Devadoss, P.R. (2010), "RFID and Organizational Transformation in the National Library Board of Singapore," *World Scientific Books*, pp. 1–26.

EPCGlobal. www.epcglobalinc.org.

Gale, T., Rajamani, D., Reyes, P.M., and Sriskandarajah, C. (2010 forthcoming), "The Impact of RFID on Supply Chain Performance," *Technology, Operations and Management*, 2(1).

Gambon, J. (2006), "RFID Frees Up Patients Beds," *RFID Journal*, Retrieved from http://www.rfidjournal.com/article/purchase/2549.

Gaukler, G., Seifert, R., and Hausman, W. (2007), "Item Level RFID in the Retail Supply Chain," *Production and Operations Management*, 16(1):65–76.

Godon, D. and Visich, J.K. (2007), "An Exploratory Study of RFID Implementation Benefits and Challenges in the Supply Chain," *Proceedings of the 38th Annual Meeting of the Decision Sciences Institute*, Phoenix, AZ, pp. 5261–5266.

Greengard, S. (2006). "Mississippi Blood Services Banks on RFID," *RFID Journal*, Retrieved from http://www.rfidjournal.com/article/view/2472.

Hardgrave, B. and Miller, R. (2006), "The Myths and Realities of RFID," *International Journal of Global Logistics and Supply Chain Management*, 1(1):1–16.

Hardgrave, B., Aloysius, J., Goyal, S., and Spencer, J. (2008), "Does RFID Improve Inventory Accuracy? A Preliminary Analysis," *Information Technology Research Institute*, Sam M. Walton College of Business, University of Arkansas, Fayetteville, AR, Working Paper ITRI-WP107-0311, available at: www.itri.uark.edu.

Hardgrave, B., Langford, S., Waller, M., and Miller, R. (2008), "Measuring the Impact of RFID on Out of Stocks at Wal-Mart," *MIS Quarterly Executive*, 7(4):181–192.

Harris, D.B. (1960), "Radio Transmission Systems with Modulatable Passive Responder," US Patent 2,927,321.

Heese, H. (2007), "Inventory Record Accuracy, Double Marginalization and RFID Adoption," *Production and Operations Management*, 16(5):542–553.

Jaska, P. and Reyes, P.M. (2007), "Better Service for Customers with RFID," *38th Annual Meeting of the Decision Sciences Institute*, Phoenix, Arizona, November 17–20.

Jones, P., Clarke-Hill, C., Shears, P., Comfort, D., and Hillier, D. (2004), "Radio Frequency Identification in the UK: Opportunities and Challenges," *International Journal of Retail and Distribution Management*, 32(3):164–171.

Kärkkäinen, M. (2003), "Increasing Efficiency in the Supply Chain for Short Shelf Life Goods Using RFID Tagging," *International Journal of Retail and Distribution Management*, 31(10):529–536.

Kärkkäinen, M. and Holmström, J. (2002), "Wireless Product Identification: Enabler for Handling Efficiency, Customization and Information Sharing," *Supply Chain Management: An International Journal*, 7(4):242–252.

Klensch, R. (1975), "Electronic Identification System," US Patent 3,914,762.

Lee, H. and Ozer, O. (2007), "Unlocking the Value of RFID," *Production and Operations Management*, 6(1):40–64.

Li, S. and Visich, J.K. (2006), "Radio Frequency Identification: Supply Chain Impact and Implementation Challenges," *International Journal of Integrated Supply Management*, 2(4):407–424.

Li, S., Visich, J.K., Khumawala, B.M., and Zhang, C. (2006), "Radio Frequency Identification Technology: Applications, Technical Challenges and Strategies," *Sensor Review*, 26(3):193–202.

Lin, C.Y. (2008), "Factors Affecting the Adoption of Radio Frequency Identification Technology by Logistics Services Providers: An Empirical Study," *International Journal of Management*, 25(3):488–499.

Matta, V., Berisso, K., and Brokaw, T. (2009), "Prototyping for the Holy Grail of RFID: Return on Investments," *Issues in Information Systems*, 10(2):536–543.

Moeeni, F. (2006), "From Light Frequency Identification (LFID), to Radio Frequency Identification (RFID), in the Supply Chain," *Decision Line*, 37(3):8–13.

O'Connor, M.C. (2009), "Alliance, Seeonic, UPM Raflatac Collaborate on Item-Level Retail Display," *RFID Journal*, Retrieved from http://www.rfidjournal.com/article/view/4820.

O'Connor, M.C. (2009), "Gillette Fuses RFID with Product Launch," *RFID Journal*, Retrieved from http://www.rfidjournal.com/article/articleview/2222/1/1/.

O'Connor, M.C. (2008), "N.C. Hospital Looks to RadarFind to Improve Asset Visibility," *RFID Journal*, Retrieved from http://www.rfidjournal.com/article/articleview/3878/1/1/.

O'Connor, M.C. (2007), "Army Medical Center Looking to Boost Asset Awareness," *RFID Journal*, Retrieved from http://www.rfidjournal.com/article/articleview/3887/1/1/.

O'Connor, M.C. (2007), "Emory Healthcare Tracks Its Pumps," *RFID Journal*, Retrieved from http://www.rfidjournal.com/article/articleview/3311/1/1/.

O'Connor, M.C. (2006), "Pro-X Seeks RFID for Internal Benefits," *RFID Journal*, Retrieved from http://www.rfidjournal.com/article/articleview/2188/1/1/.

Ozelkan, E.C. (2007), "Feasibility of RFID Decisions for Managing Supply Chains – A Returns Analysis," Invited speaker: *3rd Annual RFID-Integrated Supply Chains Symposium*, Baylor University's Hankamer School of Business, Waco, Texas, U.S.A. September 27–28.

Ozelkan, E.C. and Galambose, A. (2008), "When Does RFID Make Business Sense for Managing Supply Chains?" *International Journal of Information Systems and Supply Chain Management*, 1(1):15–47.

Ozelkan, E.C. and Galambosi, A. (2007), "When Does RFID Make Business Sense for Managing Supply Chains?" *18th Annual Conference of the Production and Operations Management Society*, Dallas, Texas, May 4–7.

Pålsson, H. (2008), "Using RFID Technology Captured Data to Control Material Flows," *Proceedings of the POMS 9th Annual Conference*, La Jolla, CA, May.

Pålsson, H. (2007), "Participant Observation in Logistics Research: Experiences from an RFID Implementation Study," *International Journal of Physical Distribution and Logistics Management*, 37(2):148–163.

Pei, J. and Klabjan, D. (2010), "Inventory Control in Serial Systems under Radio Frequency Identification," *International Journal of Production Economics*, 123(1):118–136.

Pisello, T. (2006), "The ROI of RFID in the Supply Chain," *RFID Journal*, Retrieved from http://www.rfidjournal.com/article/articleview/2602/1/2/.

Prater, E., Frazier, G., and Reyes, P.M. (2005), "Future Impacts of RFID on E-Supply Chains in Grocery Retailing," *Supply Chain Management: An International Journal*, 10(2):134–142.

Reyes, P.M. (2008), "Impact of RFID Technology on the Bullwhip Effect," *Institute for Operations Research and the Management Sciences Annual Meeting*, Washington D.C., October 12–15.

Reyes, P.M. (2008), "RFID Attractiveness in Supply Chain Management: An Update," *18th Annual Conference of the Production and Operations Management Society*, La Jolla, California, May 9–12.

Reyes, P.M. (2007), "Impact of RFID-Point of Sale Data Sharing: An Experimental Study," *18th Annual Conference of the Production and Operations Management Society*, Dallas, Texas, May 4–7.

Reyes, P.M. (2007), "Radio Frequency Identification (RFID) Implementation Efforts at Four Firms: Integrating Lessons Learned and RFID-specific Survey," *Sloan Industry Studies Annual Meeting*, Cambridge, Massachusetts, April 26–27.

Reyes, P.M. (2006), "Connecting RFID Technologies Supply Chain Operations," *37th Annual Meeting of the Decision Sciences Institute*, San Antonio, Texas, November 18–21.

Reyes, P.M. (2006), "Is RFID Marking a New Generation of Supply Chain Management?" *17th Annual Conference of the Production and Operations Management Society*, Boston, Massachusetts, April 28–May 1.

Reyes, P.M. (2006), "RFID Adoption and Implementation Issues in Healthcare Supply Chains," *37th Annual Meeting of the Decision Sciences Institute*, San Antonio, Texas, November 18–21.

Reyes, P.M. (2006), "RFID Adoption and Implementations in Supply Chains" *Institute for Operations Research and the Management Sciences Annual Meeting*, Pittsburgh, Pennsylvania, November 5–8.

Reyes, P.M. (2005), "Building a Business Case for RFID Implementation," *36th Annual Meeting of the Decision Sciences Institute*, San Francisco, California, November 19–22.

Reyes, P.M., (2005), Panelist for "RFID in the Closed-Loop Supply Chain," *36th Annual Meeting of the Decision Sciences Institute*, San Francisco, California, November 19–22.

Reyes, P.M., (2005), Panelist for "The Role of Auto-Id Technologies, RFID, Bar Code, Biometrics, and Smart Card in Quality Decision Making and Supply Chain Integration—Part 4," *36th Annual Meeting of the Decision Sciences Institute*, San Francisco, California, November 19–22.

Reyes, P.M. and Frazier, G. (2007), "Radio Frequency Identification: Past, Present and Future Business Applications," *International Journal of Integrated Supply Management*, 3(2):125–134.

Reyes, P.M. and Frazier, G.V. (2005), "RFID: Adoption Motivations and Implementation Issues," *36th Annual Meeting of the Decision Sciences Institute*, San Francisco, California, November 19–22.

Reyes, P.M. and Jaska, P. (2007), "Is RFID Right for Your Organization or Application?" *Management Research News*, 30(8):570–580.

Reyes, P.M. and Jaska, P. (2006), "A Research Agenda for RFID Integrated Supply Chain Management Studies," *International Journal of Global Logistics and Supply Chain Management*, 1(2):98–103.

Reyes, P.M., Jaska, P., and Heim, G. (2009), "Improving Customer Service with RFID Technology," *10th International Conference of the Decision Sciences Institute*, Nancy, France, June 24–27.

Reyes, P.M., Jaska, P., and Heim, G. (2009), "Improving Customer Service with RFID Technology: An Economic Analysis of Real Option Value from RFID Service Application," *Industry Studies Association Annual Conference*, Chicago, Illinois, May 28–29.

Reyes, P.M., Jaska, P., and Weeks, J. (2009), "Better Decisions Using Advanced Planning and Scheduling (APS) with RFID," *International Journal of Data Analysis and Information Systems*, 1(1):39–46.

Reyes, P.M., Li, S., and Visich, J. (2009), "Development and Validation of a Measurement Instrument for Studying RFID in the Health Care Industry," *40th Annual Meeting of the Decision Sciences Institute*, New Orleans, Louisiana, November 13–17.

Reyes, P.M., Frazier, G.V., and Prater, E. (2004), "E-Procurement and Future Impacts of RFID on E-Supply Chains in Grocery Retailing," *2nd World Conference on POM: 15th Annual Conference of the Production and Operations Management Society*, Cancun, Mexico, April 30–May 4.

Reyes, P.M., Frazier, G., Prater, E., and Cannon, A. (2007), "RFID: The State of the Union between Promise and Practice," *International Journal of Integrated Supply Management*, 3(2):192–206.

Swedberg, C. (2009), "Charles Voegele Group Finds RFID Helps It Stay Competitive," *RFID Journal*, Retrieved from http://www.rfidjournal.com/article/view/4836.

Swedberg, C. (2009), "Howard Memorial Finds RFID Keeps Assets from Getting Los," *RFID Journal*, Retrieved from http://www.rfidjournal.com/article/view/5244.

Swedberg, C. (2009), "New York Medical Center Tracks OR Equipment for Trauma Care," *RFID Journal*, Retrieved from http://www.rfidjournal.com/article/view/5013.

Swedberg, C. (2009), "RFID Illuminates Work-in-Progress for Neonlite," *RFID Journal*, Retrieved from http://www.rfidjournal.com/article/view/4691.

Swedberg, C. (2008), "Asset Tracking Underway at WakeMed Cary Hospital," *RFID Journal*, Retrieved from http://www.rfidjournal.com/article/articleview/4056/1/1/.

Swedberg, C. (2008), "RFID Helps Endwave Track Work-in-Progress." *RFID Journal*, Retrieved from http://www.rfidjournal.com/article/articleview/4168/1/1/.

Vernon, F. (1952), "Application of the Microwave Homodyne," *IRE Transactions on Antennas and Propagation AP-4*, pp. 110–116.

Vijayaraman, B. and Osyk, B. (2006), "An Empirical Study of RFID Implementation in the Warehousing Industry," *The International Journal of Logistics Management*, 17(1):6–20.

Violino, B. (2010), "Memorial Hospital Miramar Builds Benefits onto Its RTLS," *RFID Journal*, Retrieved from http://www.rfidjournal.com/article/view/7431.

Visich, J.K., Li, S., and Khumawala, B.M. (2007), "Enhancing Product Recovery Value in Closed-loop Supply Chains with RFID," *Journal of Managerial Issues*, 19(3):436–452.

Visich, J., Li, S., Khumawala, B.M., and Reyes, P.M. (2009), "Empirical Evidence of RFID Impacts on Supply Chain Performance," *International Journal of Operations and Production Management*, 29(12):1290–1315.

Visich, J., Li, S., Khumawala, B.M., and Reyes, P.M. (2008), "Empirical Evidence of RFID Impacts on Supply Chain Performance." *39th Annual Meeting of the Decision Sciences Institute*, Baltimore, Maryland, November 22–25.

Violino, B. (2009), "A Tech-Savvy Medical Organization Gives the Thumbs-Up to RFID," *RFID Journal*, Retrieved from http://www.rfidjournal.com/article/view/5172.

Wessel, R. (2009), "German Researchers to Test Networking Tags for Assets, Blood," *RFID Journal*, Retrieved from http://www.rfidjournal.com/article/view/5143.

Wessel, R. (2009), "Rewe Deploying Long-Range Real-Time Location RFID System," *RFID Journal*, Retrieved from http://www.rfidjournal.com/article/view/5187.

Wessel, R (2007), "RFID Synergy at a Netherland Hospital," *RFID Journal*, Retrieved from http://www.rfidjournal.com/article/purchase/3562.

Wyld, D. (2006), "RFID 101: The Next Big Thing for Management," *Management Research News*, 29(4):154–173.

Zaino, J. (2008), "Integris' Journey to RFID," *RFID Journal*, Retrieved from http://www.rfidjournal.com/article/view/4098.

Zane, C. and Reyes, P.M. (2010), "Airline Plight: Where Has All the Luggage Gone?" *Management Research News*, 33(7):767–782.

Zelbst, P., Green, K., Sower, V., and Reyes, P.M. (2010 forthcoming), "Impact of Supply Chain Linkages on Supply Chain Performance," *Industrial Management and Data Systems*.

Zelbst, P.J., Green, K.W., Gu, Q., and Abshire, R.D. (2010), "Impact of RFID Technology Utilization on Organizational Agility," *Industry Studies Association Conference*.

Zelbst, P.J., Green, K.W., Sower, V.E., and Reyes, P.M. (2008), "The Impact of RFID Utilization on Manufacturing Effectiveness and Efficiency within a Supply Chain Context," *RFID Conference*, Waco, Texas.

# Index

Note: Page numbers followed by *f* denote figures; page numbers followed by *t* denote tables.

## A

## F



## G

CPSIA information can be obtained at www.ICGtesting.com
Printed in the USA
LVOW05*1830210814

399860LV00001B/15/P